Development of a New Material – Monolithic Ti_4O_7 Ebonex® Ceramic

Development of a New Material – Monolithic Ti$_4$O$_7$ Ebonex® Ceramic

by

P C S Hayfield

RS•C
ROYAL SOCIETY OF CHEMISTRY

mfis

ISBN 0-85404-984-3

A catalogue record for this book is available from the British Library.

Published by: Royal Society of Chemistry, Thomas Graham House, Science Park, Milton Road, Cambridge CB4 0WF, UK Registered Charity No: 207890

and Metal Finishing Information Services Ltd, PO Box 70, Stevenage, Herts, SG1 4DF, UK

see: www.rsc.org for further information.

Typeset by Pantile Publishing, Knebworth, Herts, UK

Acknowledgements

It is a real pleasure to acknowledge the assistance rendered by the British Library whose splendid facilities are a joy to use. We are grateful to staff in the Science Reading Rooms, but especially to Vicky Tattle, Jeremy Nagle, Paul Allchin and Simon Watts.

Thanks are also due, as so many times in the past, to Jan Jackson of Pantile Publishing who typeset the book.

Contents

1. TOWARDS AN ELECTRICALLY CONDUCTIVE CERAMIC

              1

1.1 Introduction 1

1.2 The Path to a New Material 2

1.3 Initial Production of Test Material 5

2. THE PROPERTIES OF TITANIUM SUBOXIDE & ASSOCIATED PROCESSES

              9

2.1 The Titanium – Oxygen Equilibrium Diagram 9

2.2 How to Produce Ti_4O_7? 13

    2.2.1 General Considerations 13

    2.2.2 Porosity in Hydrogen-Reduced Titania Ceramic 26

2.3 Plating/Coating of Ti_4O_7 with Metals and/or Metal Oxides 31

2.4 The Electrical Resistance of the Suboxide 33

2.5 Corrosion Resistance and Electrochemical Properties 36

2.6 Electrical Connection to Monolithic Ti_4O_7 44

2.7 Handling/Machining 46

3. APPLICATIONS – I

              49

3.1 Zinc Electrowinning 49

3.2 Swimming Pool Electrochorinators 51

3.3 Cathodic Protection 58

    3.3.1 Cathodic Protection of Rebars in Concrete 59

4. APPLICATIONS – II   **67**

   4.1 Electrode Boiling 67

   4.2 Electro Osmotic Damp Proofing 69

   4.3 Chlorine and Chlorate Electrolysis 70

   4.4 Hydrochloric Acid Electrolysis 72

   4.5 Electrolytic Sterilisation of Water 72

   4.6 Other Electrochemical Applications for Monolithic Ti_4O_7 74

   4.7 Monolithic Ti_4O_7 in Lightweight Batteries 74

5. IS Ti_4O_7 AN OPTIMUM CERAMIC ELECTRODE CHOICE?   **77**

   5.1 Some General Reflections 77

   5.2 Costs and Prices 80

6. CONCLUSIONS AND EPILOGUE   **81**

   6.1 Successes and Failures 81

   6.2 Epilogue – A Gathering of Many Strands 81

      Acknowledgements 83

REFERENCES AND FURTHER READING   **85**

INDEX   **95**

1. Towards an Electrically Conductive Ceramic

1.1 INTRODUCTION

It may on first consideration seem implausible that monolithic titanium dioxide ceramic, well-known as an electrical insulator/dielectric material, could ever be transformed into an electrically conductive material, with a potential application as a commercial electrode to rank alongside existing materials such as graphite, magnetite, silicon iron and the family of noble metal/oxide-coated titanium anodes. Yet this is now the case, and what follows is a first-hand account of its inception in an industrial metallurgical laboratory, and the successes and failures, as those involved, guided largely by intuition, backed with empirical trials, felt their way towards a useful and totally novel material.

The work described, was carried out in the Central Research & Development Laboratories of Messrs IMI plc, at Witton, Birmingham, and began in 1981. As it became apparent that a novel, electrically conductive and relatively corrosion-resistant material was within reach, there were some who felt it might provide an alternative substrate to titanium as a basis for precious metal (or oxide) coated electrodes to be used in aggressive electrolytes, such as the strong sulphuric acid used in many metal electrowinning processes. However the commercial incentive for change from the widely-used lead-silver alloy anodes was weak, and it seemed to some, too big a step to introduce a brittle ceramic into an industry where large and robust components were handled with abandon.. Others contemplated an opportunity for the electrode material in reversing-polarity swimming pool electrochlorinators, a vast market. Yet others, from a manufacturing background, perceived improved methods of manufacture, leading to a much wider range of possible applications, including lower weight lead-acid-type batteries than hitherto possible. These and other applications are discussed below.

The material, for which the registered Trademark "Ebonex"™ , was secured, because of its lustrous dark grey or black appearance, has been commercially available for some years now, and has been used in a range of applications, principally cathodic protection (Fig 1-1). Originally registered to Messrs Marston Palmer Ltd (a wholly-owned subsidiary of IMI Ltd), the Trademark is now the property of Messrs Atraverda Ltd, and covers a range of electrically-conducting sub-oxides of titanium. Many other applications remain to be explored. Some are contingent on further improvement in manufacturing technique, while others, as is so often the case, have to surmount the barrier posed by existing investment in long-established plant and processes. Significant capital investment is called for before new markets can be addressed and such ventures are not without risk.

Fig. 1-1. Selection of shaped monolithic Ti$_4$O$_7$ ceramic, including rod, tube and plate.

1.2   THE PATH TO A NEW MATERIAL

A description of the events leading up to the identification of Ti$_4$O$_7$ as an electrically conductive ceramic and a possible electrode material here begins with a series of paragraphs that might seem disjointed and *non-sequitur*, but which – taken together –describe how the concept of the all-titanium oxide electrode came about.

The 1970's and 1980's were a time when many of the graphite-based anodes used in the chlor-alkali industry (including production of chlorine, chlorate and hypochlorite) were replaced by low overpotential, noble metal oxide-type, coated titanium electrodes.[36] The incentives for phasing out graphite anodes were powerful. Most chlor-alkali processes are operated at high current density meaning that quite small savings in cell voltage (typically 0.3V or so) result in substantial long-term energy savings. So rapid and rewarding was the introduction of the new coated anodes that, over and above electrode re-coating/replacement market, the electrode industry set its sights on related markets, that is on the introduction of titanium based electrodes in other processes, notably electrowinning of metals, where huge quantities of electricity were used. The anode manufacturers were confident that in these industries too, impressive cost-savings could be won. The most obvious target industries were those involved in metal winning, principally zinc, copper and manganese.

Electrowinning is, for the most part, carried out at much lower current densities than those used in chlor-alkali manufacture, and to this extent, energy savings might be expected to be smaller. The electrolyte used, however is usually strong sulphuric acid, often containing significant concentrations of hydrofluoric acid. While titanium can broadly be considered to be highly corrosion-resistant to most chlor-alkali electrolytes, the same is not true in strong sulphuric acid, even when anodically polarized. The presence of hydrofluoric acid in these electrolytes further increases corrosion rates. The result of such attack is a tendency to undermine the applied noble metal oxide electrocatalyst coatings, thereby shortening the electrode life.

The substrate attack described above, posed a dilemma for those seeking to move into these huge new markets by developing replacement anodes for the lead-silver alloys widely used in electrowinning. (For completeness, it should be noted that there have been extensive studies of other binary, ternary and quaternary lead-based alloys to replace the conventional lead silver. Some did show an apparent superiority, but for various reasons do not appear to have been widely introduced on an industrial scale). The obvious approach was to deposit coatings which would be so impervious to electrolyte as effectively to hide the substrate titanium for most of the electrode useable life. Another approach related to the use of titanium intermetallics. It had previously been noted, in pickling-off the copper used as a lubricant in certain titanium processing eg. in wire-drawing where an intermetallic forms between titanium and copper, that the last vestiges of copper were difficult to remove. When an electrochemical method using sulphuric acid electrolyte was tried, it was observed that, when the surface was anodically polarised, oxygen evolution occurred in preference to copper dissolution. The balance between the two reactions was about 98 to 2. Such intermetallics were never seriously considered as a coating for titanium substrates. However, lumps of titanium-copper intermetallic contained within a titanium mesh anode basket seemed to work satisfactorily, and these and other possibilities for titanium intermetallic-type anodes were extensively examined.

It was during the testing of such copper-titanium intermetallics that it was observed particles became covered with a titanium oxide film. This could simply have been a consequence of the selective dissolution of copper. However, a view at the time, later shown to be erroneous, was that perhaps the titanium oxide film could be controlling the rate of copper dissolution. Whether logical or not, could a form of titanium oxide be made into an electrode its own right?

As chance would have it, the company in which the above work was being progressed, had once owned a ceramics manufacturing facility. Their product range included titanium dioxide, usually in the form of tube, used as a dielectric for manufacture of capacitors. Over the years another use of this material had developed, stemming from its toughness, which allowed its use as a thread guide in the textile industry. Continuous rubbing of thread leads to build up of static electricity which in turn attracts 'dirt' which can then cause damage to the thread. Over the years it had been found that build-up of static could be diminished or eliminated by introducing a measure of electrical conduction into surface layers. This was achieved by heating the vitrified products in hydrogen at about 600°C for a few hours. The treatment caused the titanium dioxide ceramic to darken or go a blue-black, a process know as 'blueing'. As it happened, in connection with an earlier investigation into the metallising of plastics., two rods of such 'blued' titania were to hand, approx. 10mm diameter × 120mm long.

With hindsight, and given the perceived importance at the time, of developing possible alternative anodes to replace lead-silver in electrowinning, in particular of zinc, it might have seemed logical to test the 'blued' titania rods in a simulated zinc winning electrolyte. But as it happened, a simulated manganese winning solution was readily available, based on sulphuric acid and operated at the unusually high temperature of ca. 95°C. Thus it was that the 'blued' rods were tested as anodes in this electrolyte, and in the short term at least, appeared to pass current satisfactory. So, for the first time, the concept of a monolithic titanium sub-oxide electrode was demonstrated.

The concept of such an electrode could easily have been overlooked, had it not been for a succession of fortunate events. First there was the felicitous (albeit erroneous) conception of an outer film, forming on anodically polarized titanium-copper intermetallic, which displayed semiconducting behaviour, thereby providing a rate controlling effect on the anodic processes occurring at the surface. Secondly was the ready availability of some of the 'blued' titania rods; and finally there was to hand a supply of manganese electrowinning electrolyte. As was later appreciated, the manganese-winning electrolyte is much more favourable to the operation of such anodes than, for example, zinc-winning electrolytes, allowing even uncoated titanium to function satisfactorily. Had zinc-winning electrolytes been used, as might so easily have been the case, less encouraging results could have led to the project being abandoned. It was widely accepted that the IMI Research and Development Department had been wisely organised in such a way as to encourage lateral thinking. But even so, the path from initial concept to eventual commercialisation of a product and identification of appropriate markets, was to take another 20 years and to involve costs of many millions of pounds.

1.3 INITIAL PRODUCTION OF TEST MATERIAL

It was recognised at the time that the electrical conductivity of the "blued" titania was only superficial. On fracturing, the bulk of the ceramic is creamy white in colour, as is its 'non blued' precursor. It was loosely thought that the outer layer was some form of non-stoicheometric rutile , referred to as TiO_x where "x" was not known. (Full details of the titanium – oxygen equilibrium diagram had not been established at that time). Most of those working on the project had no knowledge of oxide phases, either by name, such as "Magneli" or "Wadsley" or by their stoicheometry, such as Ti_4O_7

To explore further the possibilities of a modified titanium dioxide ceramic as an electrode, it was necessary to establish a supply source, for experimentation and trials. Mainstream ceramic manufacturers were not interested in supplying such small quantities. The alternative was to identify stockists who might be prepared to do so. From such stockists, a few tubes were procured, all approximately 10mm OD $\times$ 100 mm long.

Within the Research & Development Department, equipment was available for the heat treatment of metals and alloys in non-oxidising atmospheres, including hydrogen. The commercially procured tubes were therefore heated in hydrogen at ca. 1000°C for a few hours. The 'blued' tubes which emerged, were darkened not only superficially, but throughout the bulk of the ceramic. The ceramic had become integrally electrically conductive, though with a conductivity still far short of that characteristic of most metals.

Coated with an established noble metal/oxide electrocatalyst, $Pt + IrO_x$ by a paint/thermal decomposition route (described below), the tubes were anodically polarised in simulated zinc winning electrolyte (165 gm/litre sulphuric and + 115 ppm chloride + 5 ppm fluoride). Most gave good performance, some for up to a year when, for commercial reasons, the project was halted.

In other work, an uncoated 'blued' tube was made an anode in saturated brine at ambient temperature , in order to observe its current carrying capacity. Almost immediately after switch-on, and to the amazement of the onlookers, the 'blued' material visibly disintegrated, with a pile of fine black powder forming on the floor of the tank, directly underneath. This powder was later identified as titanium oxide, and in subsequent spectrographic analysis, the starting electrode was found to have contained significant quantities of aluminium.

With hindsight, the incredulity of those involved in the experiment, that alumina was present in a nominally all-titania tube, might seem naive. The truth was that these were largely metallurgists either by training or experience, and with little knowledge of either the theory or industrial practices in the ceramics industry. Clearly, there was an urgent need to learn in some detail, what was involved in the manufacture of ceramics. Thus it was appreciated that much difficulty attends the vitrification of high melting point ceramic particles. In many instances, and this includes the manufacture of titania, a low melting point vitrifying agent is added, effectively to glue together the high melting point particles. These are sometimes known as "sintering aids". In the case of titania, such agents include clays and tungsten oxide. Such clays (for example bentonite) contain aluminium. Vitrifying agents are routinely added, whether the end use is as a dielectric material or as threadguides. Of the handful of titania tubes acquired from a stockist, half contained several percent of aluminium, and the others tungsten (both in the form of their oxides).

Asked to supply titania tubes without sintering aids, manufacturers warned that it would be difficult to ensure a dense product. They also asked what purity was required for an electrode-grade ceramic. Not really knowing how to answer at that stage, it seemed reasonable to specify impurity levels similar to those set for commercial grade titanium to ASTM Grade I. In this specification, aluminium, iron and several other impurities must each not exceed 0.05 wt%. Achieving such low levels proved far from simple. Most ceramic manufacturers offering titania, are primarily alumina producers. Furthermore, most of the equipment used in powder processing or handling is ferrous-based. Impurity levels found in initially used samples from manufacturers are listed in Table 1-1.

Many more lessons in the technology of ceramics manufacture were learned through hard experience, mainly those related to the processing of the precursor powders into vitrified ceramics. It is essential, in carrying out this transformation, to ensure a predictable shrinkage rate on vitrification. Unless this is the case, components cannot be manufactured to any reasonable dimensional tolerances. In essence, this implies production of a starting powder, ('the green mix'), in which the packing of powder particles prior to vitrification, is reproducible. It may be necessary to mix several raw materials which are usually milled, sometimes for hours or even days, in order to achieve a reproducible shrinkage rate. At some point in the processing a wax-like organic binder is added to provide mechanical strength when the product is shaped (the green state) prior to vitrification. It is also normal, once the product is shaped, to burn out this organic binder prior to vitrification.

In co-operation with one particular ceramic manufacturer (Morgan Matroc) agreement was reached that, for so called 'electrode ceramic', the 'green' state product would be analysed to ensure the minimum of impurities prior to vitrification. Should the material fail the required impurity specification, as it sometimes did, then the 'mix' could be diverted to the manufacture of dielectrics and threadguides, where the specification for impurity content was less stringent. Had the only option been to discard those batches with too much impurity, the cost could well have been prohibitive.

TABLE 1-1

Emission spectrographic analyses on a range of Titanium Dioxide Ceramics and Powders

Element	A	B	C	D	E	F
Copper	<0.01	<0.01	0.01/0.05	<0.01	0.01	<0.01
Zinc	<0.05	→				
Tin	<0.01	→				
Lead	<0.01	→				
Iron	0.01	0.01	0.01/0.05	0.05	0.05	0.05
Nickel	<0.01	→				
Manganese	<0.01	<0.01	<0.01	0.01/0.05	0.01	<0.01
Aluminium	✓	✓	<0.01	✓	✓	✓
Magnesium	0.01	0.05	<0.01	0.05	<0.01	<0.01
Cadmium	<0.01 →					
Silver	<0.01 →					
Antimony	<0.01 →					
Bismuth	not detected in any →					
Arsenic	<0.01→					
Silicon	0.01	0.01	<0.01	✓	0.05/0.1	0.01
Chromium	<0.01	<0.01	<0.01	0.01	<0.01	<0.01
Cobalt	0.01	<0.01	<0.01	0.01	0.01	<0.01
Titanium	major element →					
Zirconium	0.01/0.05	<0.01	0.01	0.05	<0.01	0.05
Molybdenum	<0.05 →					
Tungsten	<0.05 →					
Vanadium	<0.05 →					
Hafnium	<0.05 →					
Niobium	<0.05 →					
Calcium	0.05 →					
Barium	0.05	<0.01	<0.01	0.01/0.05	<0.01	<0.01
Tantalum	<0.05	<0.05	<0.05	✓	<0.05	<0.05
Boron	<0.01 →					

A) TAMM HG. B) NL Industries 20/10. C) Kronos PDR. D) USA Co + 2½% Ta.
E) Control. F) LCC French Source

✓ indicates a minor addition.

Using nominally pure titania in the 'mix', ie. without vitrifying addition, material was processed into tile shape (150 × 150mm × 3mm). Even at a vitrifying temperature 100°C degrees above normal, it was not possible to achieve a dense product. Use of a higher temperature still, risked grain growth, which would have rendered an already brittle material even more so. With a theoretical density for rutile of 4.3g/cm³, the bulk density of these tiles was ~3.9 to 4.0 g/cm³. Thus the titania for subsequent hydrogen reduction was porous. Once again, a failure to achieve what had been sought, proved felicitous.. Had it not been for this initial porosity, it would have been much more difficult than it was to achieve homogeneous reduction of the titania to a lower oxidation state.

Armed with precursor titanium dioxide of higher purity, and after heat treatment in hydrogen to produce 'blueing', the material was again tested as an anode in brine. The rapid disintegration previously observed did not occur. Evidently the aluminium-rich material binding the rutile particles together in the earlier tubes was simply dissolving in brine, as might have been expected. Interestingly, the coated tubes, although they contained alumina, on evaluation in simulated zinc-winning electrolyte remained intact, presumably because the alumina-rich areas could not be further oxidised and so remained stable.

Whether 'blued' titania was made with or without vitrifying additions, the appearance after hydrogen reduction was much the same. Hence to avoid confusion, all subsequent titania starting ware was processed without vitrifying additions. With hindsight, it is possible that this was an over-reaction and it is now accepted that in certain situations, it is advantageous to add sintering aids to Ti_4O_7 powder, in order to form a dense product for use in chloride-free aqueous environments.

In order to appreciate the course of subsequent developments, it may be helpful to describe aspects of present-day knowledge of the titanium oxygen equilibrium diagram. This is set out below.

2. The Properties of Titanium Suboxide & Associated Processes

2.1 THE TITANIUM – OXYGEN EQUILIBRIUM DIAGRAM

Figure 2-1, assembled on the basis of many contributions to the literature, illustrates the titanium-oxgyen equilibrium diagram. There are many oxides, (see Table 2-1). Principal X-ray diffraction lines for several of the oxides are shown in Fig. 2-2.

In order better to appreciate aspects of this equilibrium diagram, an attempt will be made to describe the thinking about titanium and oxygen by those working on titanium at the time. Of some, it might be said that it was a permanent pre-occupation.

It had long been appreciated that titanium is in fact a most reactive metal, that owes its ambient temperature stability, in common with aluminium and several other "valve metals", to the instantaneous formation on exposure to air, of a thin protective oxide film. In the case of titanium this film is presumed to be amorphous or a finely crystalline form of rutile. Using ellipsometry, the thickness of such films was found to be to be approx. 5nm if formed in slightly moist air, and a little thicker when formed in dry air.[88]

In the early days of titanium manufacture in the UK, much effort was invested into ways of thickening the surface oxide in order to facilitate processing e.g wire drawing. Apart from thermal oxidation, which results in oxygen absorption into the bulk titanium as well as thickening of the surface oxide film, there were other approaches for forming thicker oxide films, in some cases to several microns, by purely chemical/electrochemical methods. At one stage there was call for a 'black' surface film, thought to be of possible importance for cathodes in metal winning, where stripping of strongly adherent cathodically deposited metals can be a problem. Another potential application was for camera and other optical components, and also for purely aesthetic reasons. Several techniques for forming such black surfaces were identified including anodising in alkaline solution containing sodium gluconate.

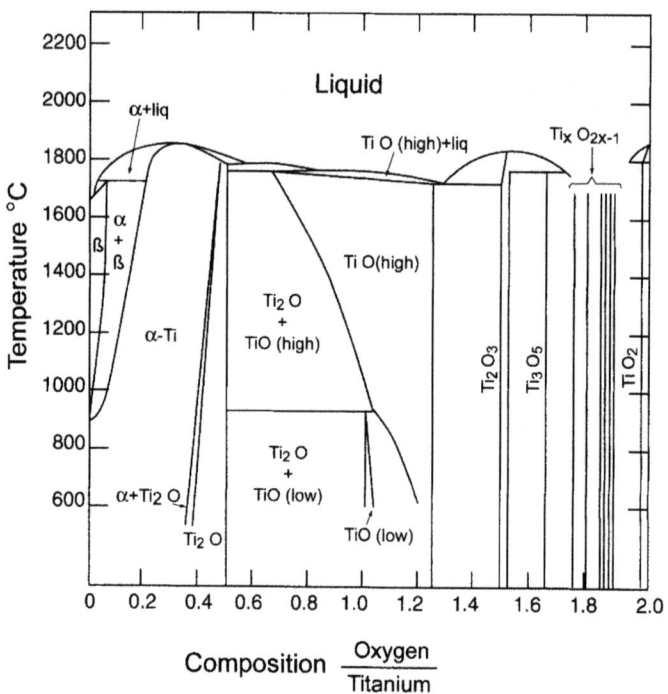

Fig. 2-1. Titanium-oxygen equilibrium diagram.

Another purely chemical method of blackening involved immersion in sulphuric/chromic acid solution at ~140° C. It was loosely thought that conditions for a good black were intermediate between 'passivation and corrosion', thereby implying quite specific conditions for its formation. The black appearance was thought to be based on optical absorption rather than light scattering, and was attributed to non-stoicheometry in the titanium oxide formed at the surface, loosely designated as TiO_x where x was unknown.

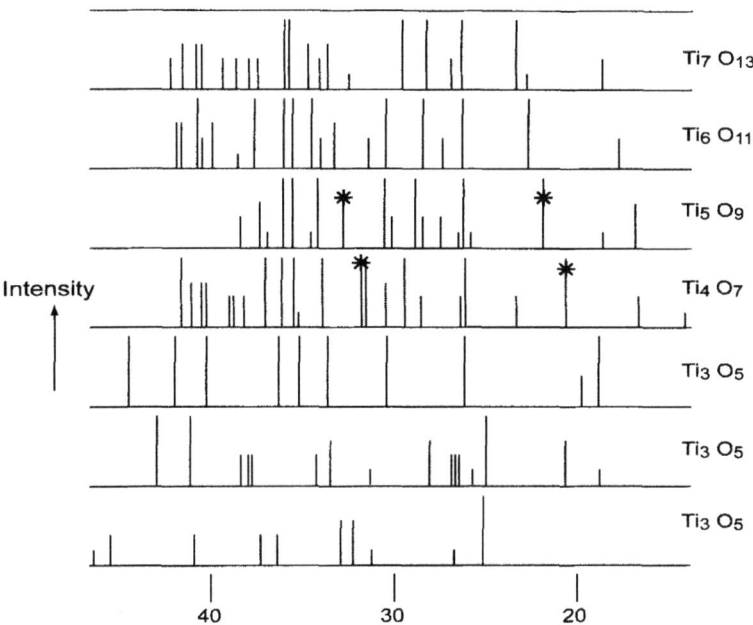

* Helpful lines to aid in the recognition of Ti₄ O₇ and Ti₅ O₉ in mixtures of oxides

$Ti_4 O_7$ 2θ   20.7°   and   31.7°
$Ti_5 O_9$ 2θ   21.9°   and   33.5°

Fig. 2-2. X-Ray diffraction spacings for selected oxides of titanium.

TABLE 2-1   Information on Titanium and its Oxides

Compound	x in TiO_x	Structure	X-ray density
TiO_2	2	Rutile	4.25
		Anatase	3.89
$Ti_{10} O_{19}$	1.9		
$Ti_9 O_{17}$	1.89	Triclinic	3.75
$Ti_8 O_{15}$	1.875	Triclinic	3.84
$Ti_7 O_{13}$	1.857	Triclinic	3.9
$Ti_6 O_{11}$	1.833	Triclinic	4.0
$Ti_5 O_9$	1.8	Triclinic	4.31
$Ti_4 O_7$	1.75	Triclinic	4.32
$\gamma\, Ti_3 O_5$	1.67	Monoclinic	4.35
		Monoclinic	4.24
		Monoclinic	4.11
$Ti_2 O_3$	1.5	Tetragonal	4.585
TiO	1.0	Hexagonal	5.69
		Cubic	5.82
		Monoclinic	5.89
$Ti_2 O$	0.5	Hexagonal	5.0
Ti	0	Hexagonal	4.5

It was for a long time thought that, once such a surface oxide film had formed it was "permanent", i.e chemically inert, and for the most part, especially in gaseous environments, this is by and large true.

When immersed in an electrolyte, it can be quite misleading to suggest that such films are inert, and studies of these black films provided early evidence of this. Ellipsometry proved a powerful technique for in-situ study of the behaviour of such black oxide films as their environment was altered. Thus a film may be formed electrolytically to reach a thickness of many tens of nm in, for example, neutral electrolyte. However if the temperature of the electrolyte is increased, or the pH lowered, the film dissolves, to be re-formed as one with greater optical absorbtion, presumably reflecting a metastable equilibrium with the changed environment. Such observations can be made either by changing the electrolyte conditions in-situ, or by removing a sample of blackened titanium from one solution, and placing it in a different solution, in each case carrying out the ellipsometric observations. Similar information has been obtained using radiotracer methods to study other systems, such as iron. At the time such ellipsometric investigatory work was being pursued, the information shown in Fig. 2-1 was not known, and compositional analysis of such films was not possible to the accuracies allowed using more modern X-ray diffraction equipment. Whether it would ever be possible to directly form Ti$_4$O$_7$ by electrolytic means, remains an interesting speculation.

At the titanium-rich end of the titanium-oxygen equilibrium diagram, it has long been known that oxygen is soluble in titanium, as are nitrogen and carbon. Oxygen in the lattice increases hardness and this forms the basis of a range of harder, commercially-available titanium materials .

A German manufacturer (Conradty), whose product range was largely based on anode-grade graphite for the chlor-alkali; industry, met the challenge of the new titanium-based electrodes by adopting powder metallurgical routes to titanium products. In the early stages of change in the chlor-alkali industry, it was difficult to make durable titanium-based anodes for the mercury-type plants widely used in Europe. The main problem was believed to occur when the liquid mercury, flowing down the trough, suddenly surged. This would result in direct metal-to-metal contact between the mercury (cathode) and the anode, in other words setting up a total short-circuit at currents of tens of thousands of Amperes. The effect was attributed to formation of so-called 'mercury butter', when the concentration of sodium dissolved in the mercury begins to lead to solidification of the mercury. To limit such damage, Conradty found it advantageous to hide the applied noble metal/ oxide electrocatalyst coating beneath a thick, porous oxide film. The latter was formed by applying to the surface a mixture of titanium metal and TiO$_2$ particles and heating to elevated temperature in argon, allowing solid state formation of Ti$_2$O and TiO. Both these oxides are several orders of magnitude more electrically conducting than Ti$_4$O$_7$ so that even when present at several micron thickness, they did not create a significant internal IR drop, even at the high current densities used in mercury-type chlorine cells. Later, other firms were able to deposit film of similar composition, sometimes rather more complex, by spraying techniques.

Mention of the formation of Ti_2O and TiO in this application is described because from patents at the time, it might have been construed that monolithic Ti_2O and TiO could be formed into practical electrodes in their own right, but it is not known whether the concept was ever brought to fruition.

Early thinking on the composition of non-stoicheometric rutile form of titanium oxide envisaged this as an n-type semiconductor, with oxygen vacancies randomly sited within the lattice (see Hauffe also Kubaschewski and Hopkins.[26,27]) As progressively more oxygen atoms are removed from the lattice, so the number of vacancies increases. The idea was discounted on thermodynamic grounds, that these vacancies could remain randomly located. Instead, the preferred model invoked vacancies collecting into planes, with ultimate closing of the lattice either side, i.e a partial collapse. The implication is of planes of titanium ions being closer to each other than might otherwise have occurred. Such defects are known as Wadsley defects.[40] As oxygen is progressively removed, more Wadsley defects occur until it is energetically more stable for the lattice to rearrange into a different symmetry from that of rutile. The structure existing at maximum Wadsley defect concentration is Ti_4O_7, and the complete restructing results in the composition Ti_3O_5. The various crystal modification up to Ti_4O_7 are known as the Magneli series $Ti_xO_{(2x-1)}$ where x ranges from 4 to 20.

Had the details of the titanium-oxygen equilibrium diagram as it is known today, been available at the time of the first experimentation to form a monolithic titanium oxide electrode, and the composition Ti_4O_7 known to be optimal (see later), then it is doubtful any further work would have been undertaken. The compositional range for Ti_4O_7 ($TiO_{1.75}$) being so narrow, it would have seemed impractical to try to manufacture such material on a commercial scale. But of course experimentation had preceded a detailed understanding, and showed that indeed it was possible to make pure Ti_4O_7 without too much difficulty.

By definition, equilibrium diagrams do not predict kinetics. The most stable oxide in the titanium-oxygen system is TiO_2, with other lower oxides in effect being metastable. Thus it is likely that all sub-oxides, once exposed to air, become covered with a surface film albeit perhaps no more then a few atomic layers in thickness, of TiO_2; subsequent experimentation which confirmed this, is described below.

2.2 HOW TO PRODUCE Ti_4O_7?

2.2.1   General Considerations

From thermodynamic considerations, (see the temperature – composition equilibrium diagrams), titanium dioxide cannot be reduced directly to the metal via reductants such as hydrogen, carbon and carbon monoxide. At sufficiently elevated temperature, however, it is known such reductants can be used to make Ti$_3$O$_5$, possibly Ti$_2$O$_3$, but certainly down to a lower oxidation state than the Magneli series, including Ti$_4$O$_7$.

Because of the immediate availability of equipment, the first reduction of monolithic titania to Ti$_4$O$_7$ was carried out using hydrogen. The experimentation involved a Nimonic (nickel-base alloy) box as sketched in Fig. 2-3.

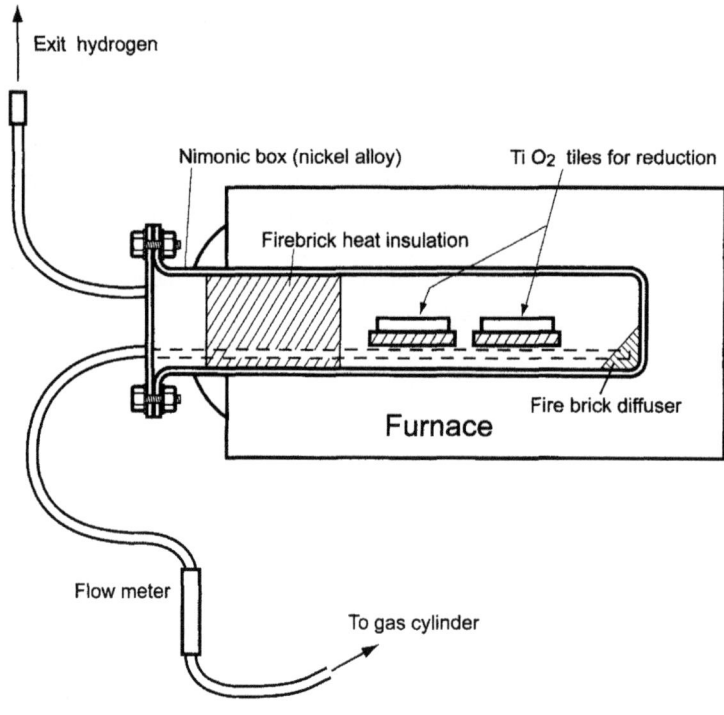

Fig. 2-3. Furnace arrangement for high temperature reduction of TiO$_2$ ceramic to monolithic Ti$_4$O$_7$.

While the furnace into which the box fitted had a capability of 1500°C, the Nimonic material began to suffer significant oxidation above 1000°C. At the minimum temperature eventually found necessary to achieve Ti$_4$O$_7$ in reasonably short times – hours rather than days – (see Fig. 2-4), the Nimonic thinned to the point of perforation over 8 to 10 heating cycles and the repair/renewal of these boxes became a significant cost.

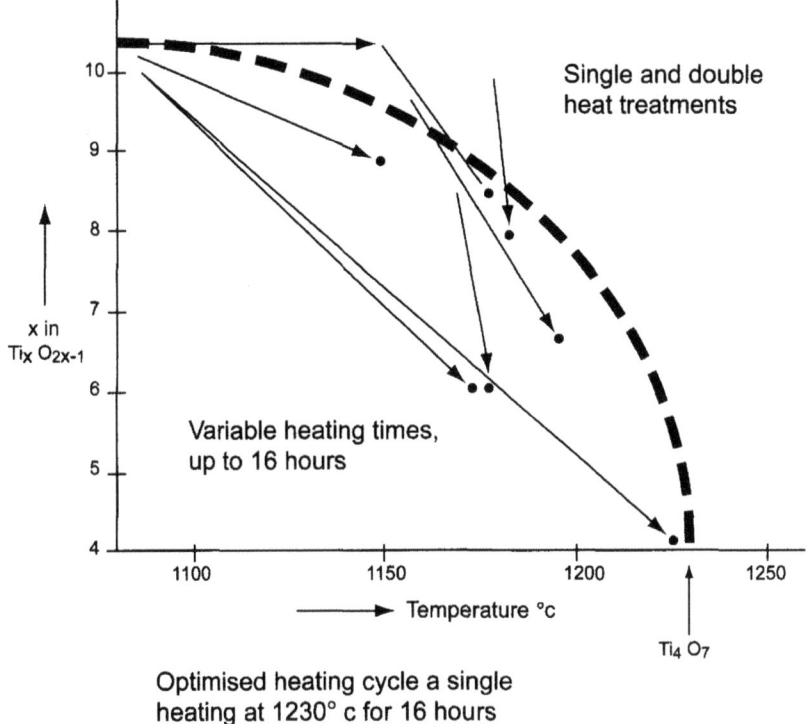

Fig. 2-4. Reduction of TiO_2 ceramic on heating in hydrogen at various temperatures.

Cylinder hydrogen was piped to the closed end of the box and into a porous fire brick to act as a kind of distributor for the gas. The titania tile for reduction was mounted on highly porous fire brick, in which large holes had additionally been drilled, as shown in Fig. 2-5. With the end of the box sealed, argon was flushed through the system to expel air, followed by hydrogen which was ignited at the outlet. Only then was the furnace switched on and the box heated to the prescribed temperature. After a specified time, the furnace was cooled down and the hydrogen replaced by argon before opening up the box.

To illustrate the reaction $4TiO_2 + H_2 = Ti_4O_7 + H_2O$, the cool end of the box and lid, upon opening up after a reduction run, were soaked in water, as one of the reaction products. In subsequent reductions, drying trains were incorporated in the outlet. Attempts were made, using these, though with little success, to measure the weight of water released during a given heat treatment cycle in order to assess the extent of reduction taking place.

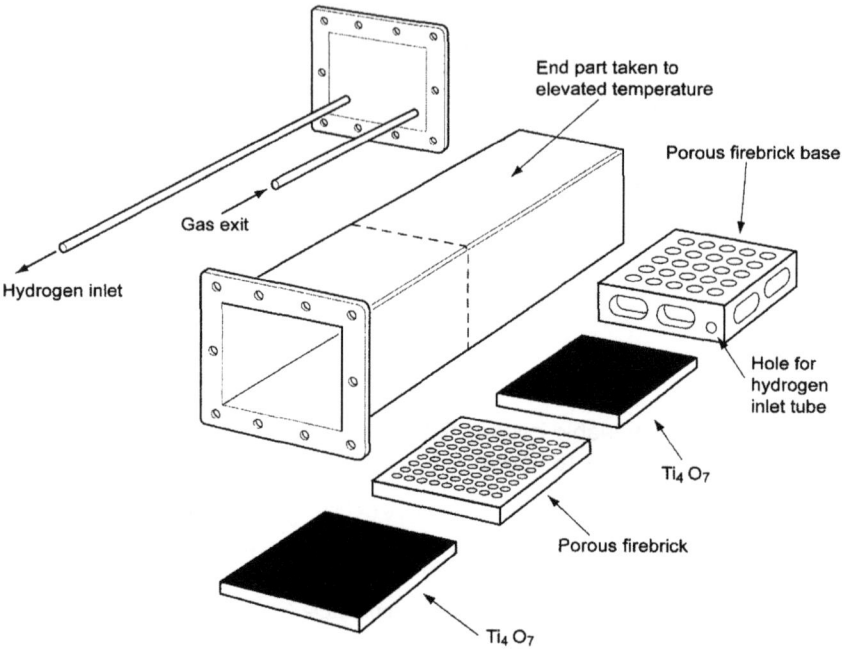

Fig. 2-5. Detail of the hydrogen reduction of TiO$_2$ ceramic to monolithic Ti$_4$O$_7$.

After a reduction run, visual examination would quickly reveal whether there had been any over-reduction to form Ti$_3$O$_5$, as the latter phase, apart from mechanically being very weak, is a pinkish in colour. Slight over-reduction near the point of hydrogen entry would result in pink spots. Gross over-reduction would mean the pinkish colour penetrating well below the surface into the bulk, and yielding material as easy to break as a biscuit.

Most reduced titania tiles had a uniform black appearance, but such visual evidence provided no indication of level of reduction or any variation in composition across the plane of the tile or in cross-section. A rapid non-destructive method of testing was to measure local surface electrical conductivity with either a multi meter or 4 point probe. Initially, until the technique had been perfected, there were sometimes alarmingly large variations across even a single tile. Another option to assess reduction level was to measure weight loss (see Fig. 2-6). This technique was not totally reliable, as once reduced tiles were exposed to air, the nature of the porosity was such that weight increased due to absorption of atmospheric water vapour, presumably by capillary action. While it was not possible to weigh to fraction of milligrams, nevertheless an order of magnitude of overall reduction could be assessed in this way. Some examples are given in Table 2-2.

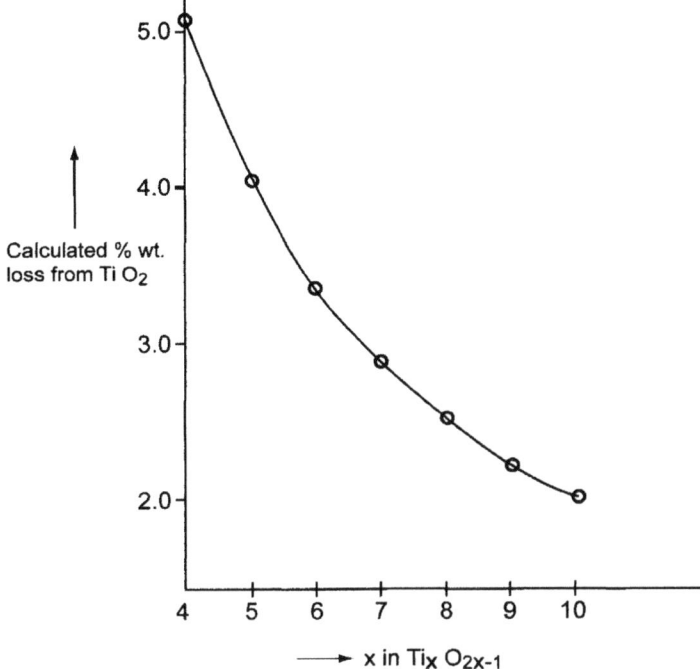

Fig. 2-6 Calculated weight loss for uniform reduction of TiO_2 ceramic to lower oxide form (complete conversion to Ti_4O_7 corresponds to weight loss of approx. ~ 5%).

TABLE 2-2

Change in density of TiO_2 tiles on reduction in hydrogen to Ti_4O_7

Each tile ~ 400 g weight. Results taken from early pilot-scale manufacture to indicate the level of consistency obtainable.

Sample Number	Initial density	Density on conversion to Ti_4O_7
EA 185	3.97	3.64
EA 186	3.97	3.63
EA 187	3.97	3.64
EA 189	3.99	3.66
EA 190	3.97	3.65
EA 191	3.95	3.67
EA 193	3.96	3.71
EA 194	3.96	3.76
EA 195	3.96	3.75
EA 196	3.99	3.79
EA 197	4.0	3.68

As already indicated, early reduction runs did not yield a uniform product, and this was especially the case for items stacked some distance from the point of entry of the hydrogen. At the time attempts were made to improve matters by increasing hydrogen flow rate and by selecting hydrogen of progressively lower initial dew point. As was later recognised, such measures proved the wrong way of achieving product uniformity.

An interesting observation at the time, reflecting the continuing metallurgical ideology that persisted, related to means of salvaging tiles with obvious spots of over-reduction, and also warped tiles which needed straightening in order to be of use. Material was replaced into the Nimonic box described above, and loaded with a heavy steel weight. Using only a slightly reducing atmosphere to avoid re-oxidation, argon with 2 to 3% hydrogen, the charge was taken up to 1230°C for a few hours. Almost miraculously, the spots of pink Ti_3O_5 had disappeared, compositional uniformity greatly improved and warping limited or reversed. Such homogenising treatments did not find favour at the time because of the cost involved in this second very high temperature heat treatment. Adequate levels of homogeneity were subsequently achieved by ensuring a more homogeneous titania starting product and ensuring that the dew point of the hydrogen during reduction remained constant and low during the reduction process.

Also shown in Fig. 2-5, is a view of the products of an early reduction run. Table 2-3 shows data for the lowering of apparent density as the reduction to Ti_4O_7 proceeds. The photograph in Fig. 2-7. illustrates a hydrogen reduction facility of greater capacity than those initially employed.

Fig. 2-7. Facility for the hydrogen reduction of TiO_2 ceramic to monolithic Ti_4O_7.

TABLE 2-3

Change in weight, volume and density of tile prepared by sintering of TiO_2 powder followed by reduction

A tile was prepared by mixing Kronos TiO_2 powder (2-4 μm particle size) with 12% Mobilcer organic binder. The dried and sieved mixture was compacted at a pressure of 1 tonne/cm². The wax was burnt off at 350°C over 18 hours. The tile was sintered 2 hrs at 1420°C. Reduction in hydrogen was carried out at 1270°C for 9½ hrs.

Stage	Weight, g	Volume, cm³	Density
After dewaxing	52.8399	25.161	2.1
After sintering	52.8399	13.959	3.78
			(+ 80% increase)
After reduction	50.8135	14.58	3.48
			(- 8% decrease)

The composition of reduced titania can be determined by powdering material and obtaining an X-ray powder diffraction diagram. Another method, albeit a destructive one, was to weigh a sample before and after re-oxidation by heating in air to 1000°C for about an hour. If A = weight of reduced tile and B = weight after re-oxidation, then x in TiO_x is given by:

$$\frac{79.9 \times (B - 16)}{B \times 16}$$

As an example, let B = 2.4876 g and A = 2.452 g, then × = 1.93.

Using the Nimonic box already described, titanium dioxide powder could be reduced to Ti_4O_7 powder. The starting material was placed in shallow ceramic trays to a depth of say 20 mm. Reduction time at 1230°C was but a few hours compared with 10 or more hours to reduce the same material in tile form. After the reduction, the product was partly fused and 'biscuit' like. It was then broken up into lumps and ball milled to a specified particle size. Any material showing signs of pinkish Ti_3O_5 colour was obviously discarded prior to milling.

To check on the electrical conductivity of Ti_4O_7 powder, material can be put onto a mould between the jaws of a press, as illustrated in Fig. 2-8. As pressure is increased, (see Fig. 2-9), so the effective electrical conductivity of the compressed powder increases dramatically.

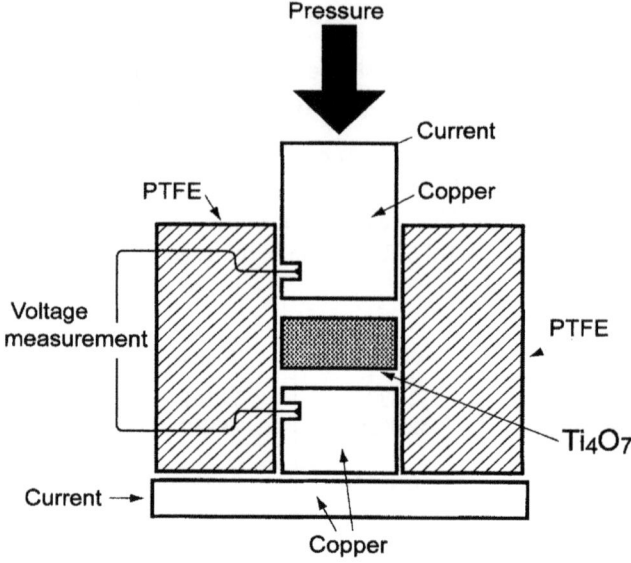

Fig. 2-8. Method for measuring electrical resistivity of Ti$_4$O$_7$ powder allowing increase of applied mechanical pressure.

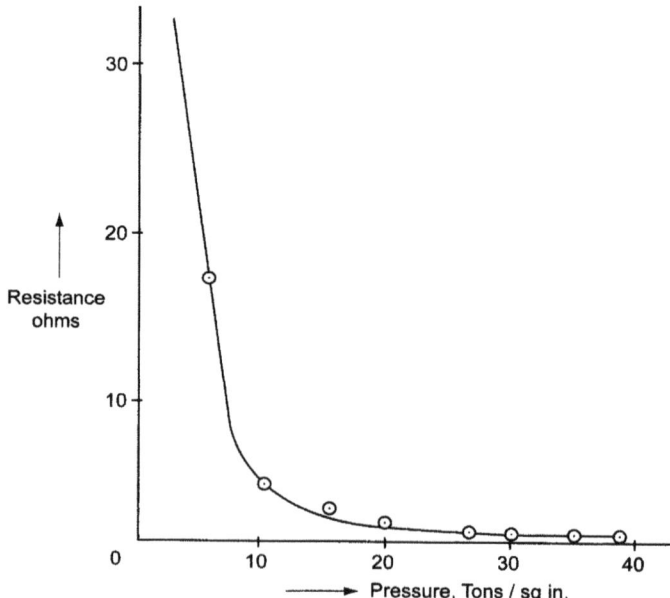

Fig. 2-9. Influence of increased mechanical pressure on the electrical resistivity of compacted Ti$_4$O$_7$ powder.

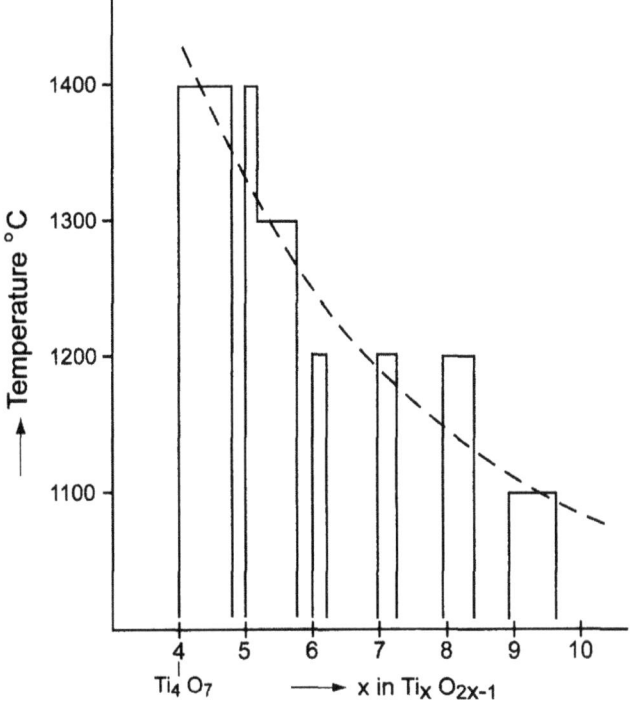

Fig. 2-10. Auto-reduction of titanium oxalate on heating in argon.

A Japanese patent suggested that TiO_2 could be reduced at relatively low reaction temperature in the presence of platinum. However a trial did not reveal any evidence for such an effect, (see Table 2-4.)

TABLE 2-4

Reaction of TiO_2 in Hydrogen in the Presence of Platinum Wire

Teachings from a Japanese patent suggested that TiO_2 could be reduced much faster in the presence of platinum than without at relatively low reaction temperature.
Rutile in powder form was heated in a refractory boat in hydrogen at 850°C for 6 hours, in one instance without the presence of platinum, and in a second case alongside a coil of platinum wire. In both instances the product was an unsintered black powder. Examined by XRD, there was no significant difference in structure, and very small reduction from the starting TiO_2.

While hydrogen reduction was used initially to form monolithic Ti$_4$O$_7$ and is still widely used, there are several other ways to form the compound, principally using carbon, (see Table 2-5). Some early production runs were carried out by a US company, then known as Lambertville Ceramics, in which carbon was successfully used. In a modified procedure, the same company patented a variant technique using graphite intercalated with one of a range of additional elements in order to improve electrical conductivity.[33]

TABLE 2-5

Miscellaneous Solid State Reactions

1.	TiO$_2$ + graphite	6 hrs at 1300°C in argon	80% Ti$_4$ + 20% Ti$_5$ + 0.2/0.4 C
2.	TiO$_2$ + graphite	6 hrs at 1230°C in argon	20% Ti$_3$, 44% Ti$_5$, 36% Ti$_6$
3.	TiO$_2$ + graphite	6 hrs at 1230°C in alternate furnace to 2.	50% Ti$_4$, 50% Ti$_3$
4.	TiO$_2$ + graphite	6 hrs at 1265°C in argon	no Ti$_4$, 30% Ti$_5$, 60% Ti$_6$ & Ti$_7$
5.	TiO$_2$ + graphite	6 hrs at 1300°C in argon	20% Ti$_4$, 70% Ti$_5$, 10% Ti$_6$ & Ti$_7$
6.	TiO$_2$ + carbon black powder To produce Ti$_3$O$_5$	6 hrs at 1300°C in argon	95% Ti$_4$ + some Ti$_3$O$_5$
7.	TiO$_2$ + carbon	6 hrs at 1300°C	90% Ti$_3$ + 10% Ti$_4$ (C 0.6%)
8.	TiO$_2$ + carbon	8 hrs at 1300°C	Ti$_3$ (C 0.16%)
9.	Ti$_3$O$_5$ + TiO$_2$	3 hrs at 1300°C in argon	Ti$_4$O$_7$ only

A wide range of alternate solid state reactions were explored as possible practical routes to making Ti$_4$O$_7$

Ti$_4$O$_7$ has been manufactured in fibre form by spinning of a viscous fluid 'mix'. .Spongy forms have also been made, by introducing a mobile 'mix' into a spongy plastic which is then burned off. Ti$_4$O$_7$ can be deposited directly onto substrates, by spraying in suitable atmosphere, using TiO$_2$ powder as starting material.

During the chemical etching of titanium in hot oxalic acid solution, it was noticed that if baths were over-used, a precipitate of titanium oxalate deposited on the sides of the holding tank. Such precipitate had a platelet form, and this gave rise to the thought it might be possible to make Ti$_4$O$_7$ in platelet form. On heating the titanium oxalate in argon it was unexpectedly found the carbon content acted as reductant to form a lower titanium oxide. By control of the time and temperature of heat treatment, a range of the lower oxides could be made, including Ti$_4$O$_7$, (see Fig. 2-10). However in the process, the platelet structure was lost.

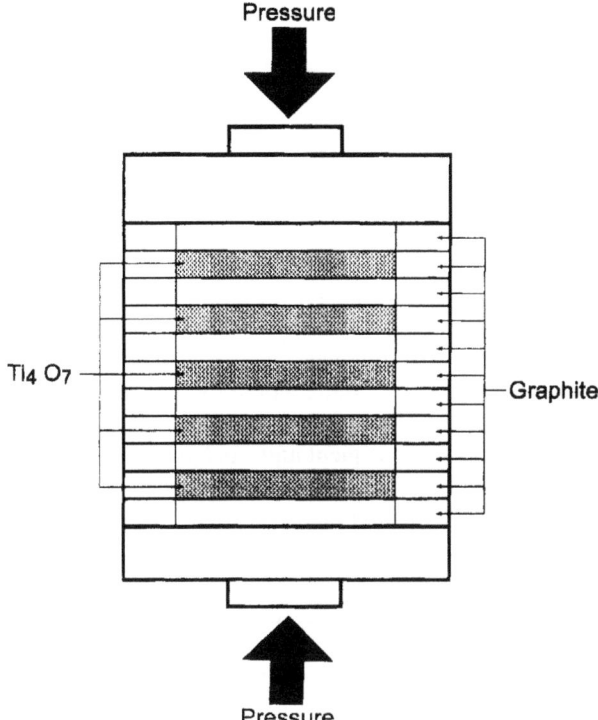

Fig. 2.11 Uniaxial hot pressing to form fully dense monolithic Ti$_4$O$_7$ plate.

The raw material used in making monolithic titania, for subsequent reduction to Ti$_4$O$_7$, is designated by the industry as 'ceramic' quality as opposed to the 'pigment' grade widely used in paint manufacture. Materials variously made throughout the world may differ slightly in composition, crystal and particle size and also particle agglomeration characteristics. Over the years, the various batches of such material received, were almost always of satisfactory purity for the purpose of electrode manufacture, although in one instance there was an unacceptably high level of antimony.

At the same period as the investigations into monolithic Ti$_4$O$_7$ as an electrode material were being progressed, a research team within ICI, including Kendall, Alford and the late Professor Derek Birchell,[80-82] were carrying out pioneering work on improving the mechanical properties and handling characteristics of ceramics and concrete. One particular development to attract widespread attention was the production of coil springs, made of concrete. Not only did the approach of the ICI team place great emphasis on the uniformity of packing of particles (where particle size and distribution were critical), but in addition, the conventional waxy organic binder was replaced by a polymer. Much of the brittleness traditionally associated with ceramics was associated with defects in the packing, which they postulated was greatly minimised by introducing shear into the mixing process. Thus the green state dough or mix was repeatedly put through asymmetric rolls to eliminate internal voidage. Applying the same techniques to starting titanium dioxide powder, it was found that on vitrification, a much tougher and denser product resulted. Such material, however, proved difficult to reduce in hydrogen to Ti$_4$O$_7$ at least uniformly so.

At one stage an expensive facility was set up to implement the ICI methodology for making dense Ti$_4$O$_7$ from starting Ti$_4$O$_7$ powder. In the event, because almost all previous methods of making Ti$_4$O$_7$ involved hydrogen reduction of a porous starting titanium dioxide, a last minute decision was to retain this approach in the interests of minimising the risk of failure. For this latter route to be successful, and despite the mixing involving shear, the monolithic titania needed deliberately to be made porous in order for subsequent uniform hydrogen reduction to Ti$_4$O$_7$. In retrospect, the decision not to process starting with Ti$_4$O$_7$ powder was an unfortunate one, and practising the ICI method on starting Ti$_4$O$_7$ remains possibly the most practical route for making dense monolithic material.

As part of a programme in which many variants of processing to make monolithic Ti$_4$O$_7$ were tried, a quantity of Ti$_4$O$_7$ powder was submitted to a commercial firm (Unilator) for vitrification under pressure. The resultant approximately 1cm^3 sample was found to be fully dense, and key parameters were measured using this small sample. However, at the time it was held that hot isostatic pressing ("HIPPING") would be too expensive for manufacture of material in volume, and the matter was taken no further. In later years, however, it became apparent that the presence of interconnecting pores in larger batches of hydrogen reduced titania was severely limiting some potential applications. Thus problems arose as a result of creeping corrosion caused by electrolyte wicking up to connection terminals, even though these were above electrolyte level. A further trial of vitrification under pressure was therefore considered. In order to make reasonably sized tiles, say 15 × 15 cm, large presses were required. The only ones appropriate were located in western USA at the then enormous cost of US $ 10,000 per day. At these costs it was imperative to fit as much material between the jaws of the press as possible, resulting in an arrangement as shown in Fig. 2-11. On the assumption that the conditions for uniaxial hot pressing were so severe that almost any mix would densify, the starting powder was not prepared with the customary thoroughness prevalent in standard ceramic practice. Certainly the concepts developed by the ICI workers of introducing shear into mixing were not carried out. After several expensive runs, it was decided that higher pressure still, was required, thereby decreasing the area of material between the jaw of the press. A fully dense product was eventually made which turned out to be extremely brittle, more glass-like than the highly porous hydrogen reduced titania. Later it was to be found that the Ti$_4$O$_7$ had re- crystallised with what was presumed was grain boundary migration of iron. This conclusion was reached in part because the material was characterised by regions of relatively high iron content which much impaired corrosion resistance, especially in solutions containing hydrofluoric acid. With the material so brittle, cracking was almost inevitable as the anode sheets were tightly bolted up to prevent leaks in various designs of electrochemical cell, and because of this, together with the impaired corrosion resistance, the idea of uniaxial hot pressing was taken no further.

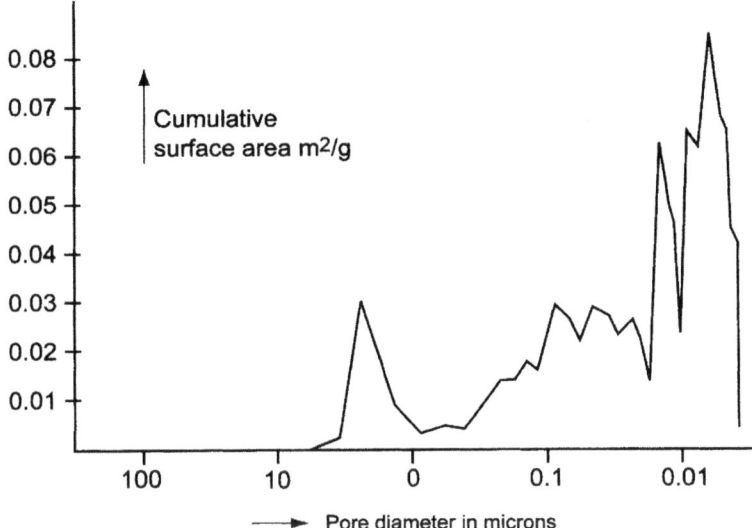

Fig. 2-12. Data on cumulative real surface area and pore diameter (from mercury porosimetry) for monolithic Ti_4O_7 tile EA27.
(Courtesy of Coulter Electronics Ltd).

With continuing difficulty in finding practical ways of making dense Ti_4O_7, attention focussed again on early investigations based on incorporation of Ti_4O_7 particles into an organic mix, such as hot pressed epoxy resins (see Table 2-6.) Inevitably in using moulding methods of this kind, electrical conductivity will suffer unless there is a very high particle content, certainly no less than 60%. A successful product has now been achieved, providing adequate electrical conductivity in material with no interconnecting porosity. Such material is now used in battery manufacture.

Looking back over the 20 or more years of trying to perfect methods of manufacture of monolithic Ti_4O_7 ceramic, and then persuading industry to manufacture it, two lasting thoughts come to mind. The first was the pronouncement, early on in the programme, by a ceramics manufacturer that, as far as ceramics are concerned, "Small is beautiful". This has certainly turned out to be the case. While tiles or plates up to foolscap (200 × 300mm) size have been made, there has never for a moment, been any suggestion, even with tape forming, that large pieces akin to standard size titanium sheets (2500 × 1000 mm), would ever be possible or practical.

Secondly, the author has been saddened, over the years of this work, by the general reluctance of UK firms to participate in new technology such as the use of monolithic ceramic electrodes for the electrochemical industry, as described here. There are honourable exceptions, of course. But it is hard to deny a general climate which appears to be risk-averse, and which lacks the enthusiasm for new ideas which is almost the norm in many other countries.

TABLE 2-6

Change in Electrical Resistivity of Hot Pressed Cylinders of Ti_4O_7 / polyvinylidene fluoride (PVDF) of varying ratios

PVDF : Ti_4O_7 ratio	Density g/cc	Electrical Resistivity Ω cm
1 : 0.6	2.07	10^7
1 : 1.9	2.53	8.9×10^3
1: 4.5	2.36	2.4×16^3
1 : 4.5	2.45	1.7×10^3
1 : 5	2.32	63
1 : 5.5	2.73	45
1 : 6	2.50	42

Pressing Conditions 180°C, 200 bar, 10 mins.
Cylinders of ~ 300 mm dia.

2.2.2   Porosity in Hydrogen-Reduced Titania Ceramic

Formation of interconnecting pores in hydrogen reduced titania has already been mentioned several times. In part this is due to porosity in the precursor monolithic titanium dioxide, but it is much increased during the hydrogen reduction to Ti_4O_7, due to successive phase changes and the removal of water vapour.

To learn more about the above the form of such porosity, samples were circulated to several laboratories specialising in characterisation of porous materials using such techniques as gas absorption and mercury porosimetry. In mercury porosimetry, mercury is forced into pores under pressure. The pore diameter penetrated is inversely proportional to mercury pressure; thus a scan of volume of mercury intruded as a function of pressure allows a pore volume/diameter distribution to be derived. Commercial equipment can be operated automatically to a maximum of 2000 atmospheres, corresponding to pore diameters down to

7.5 nm. Applied to monolithic Ti_4O_7 typical data is shown in Fig. 2-12. The information suggests the structural form illustrated in Fig. 2-13. While it is always dangerous to generalise, there would seem to be relatively large cavities of up to a few microns in size, dictated by the starting particle size and agglomeration of the TiO_2 powder. Additionally there are also finer capillaries, with diameters down to a few nanometres.

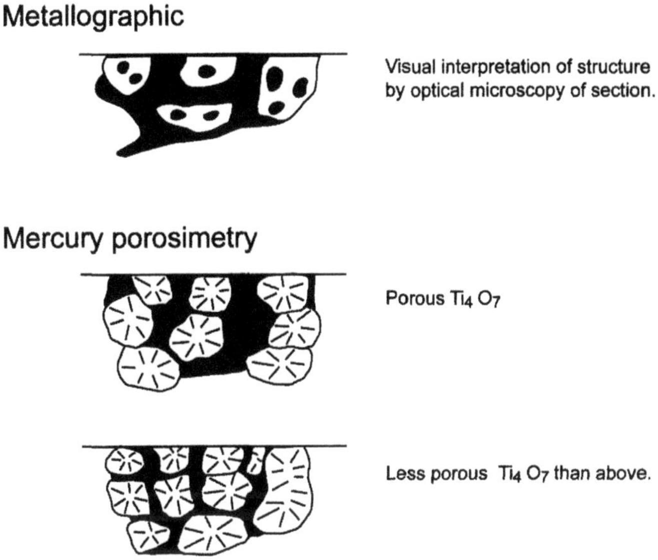

Metallographic

Visual interpretation of structure by optical microscopy of section.

Mercury porosimetry

Porous $Ti_4 O_7$

Less porous $Ti_4 O_7$ than above.

Fig. 2-13. Pore structure of monolithic Ti_4O_7 ceramic as deduced from metallographic examination and mercury porosimetry.

The latter are probably responsible for the instantaneous uptake of atmospheric water vapour on first exposing freshly reduced material to the atmosphere, as a result of capillary action. Tests of material exposed at atmospheres of varying water vapour pressure confirmed a relationship between weight uptake and water vapour partial pressure.

A quick test for the presence of interconnecting porosity is the simple drop test. A spot of water, or an alcohol such as ethanol, is dropped onto the specimen and time for absorption noted. A spot on a glass slide might stay for many hours, depending upon ambient conditions, with evaporation the only means by which the liquid could be lost. By contrast, a spot placed on freshly made Ti_4O_7 vanishes within a second, as it might on blotting paper. If spots are repeatedly applied, then liquid leaks out on the other side. Depending on the porosity of the sample, a range of intermediate behaviours can be observed.

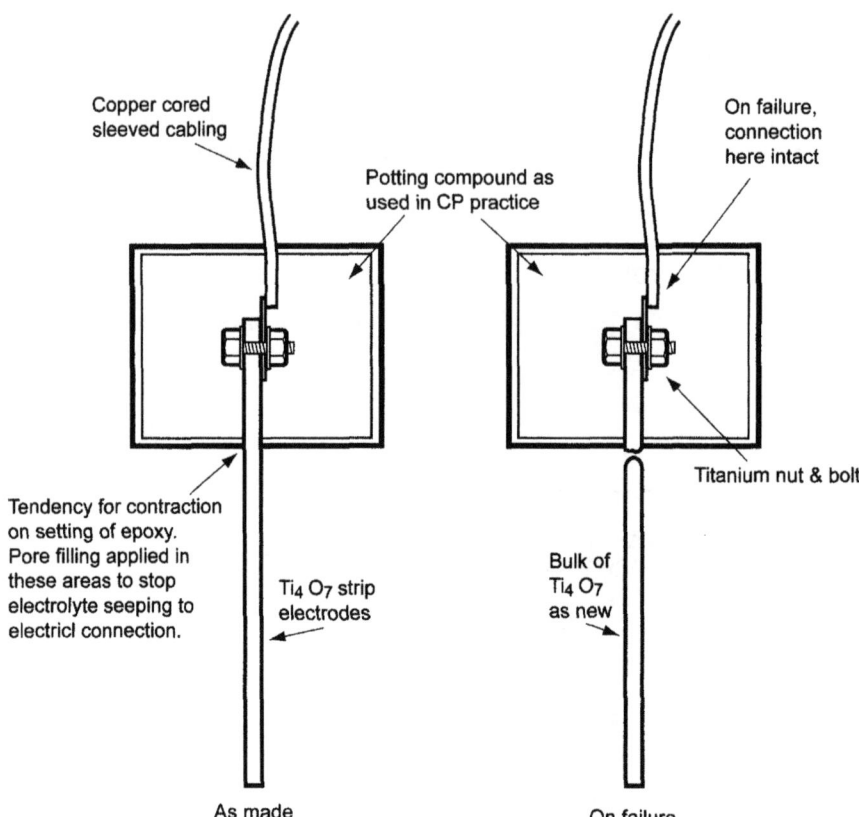

Copper cored
sleeved cabling

Potting compound as
used in CP practice

On failure,
connection
here intact

Tendency for contraction
on setting of epoxy.
Pore filling applied in
these areas to stop
electrolyte seeping to
electricl connection.

Ti$_4$ O$_7$ strip
electrodes

Bulk of
Ti$_4$ O$_7$
as new

Titanium nut & bolt

As made

On failure

Fig. 2-14. Effect of localised pore filling of monolithic Ti$_4$O$_7$ with
dichtol on its premature failure as an anode in seawater.

(Brixham Seawater Laboratory).

In retrospect, the presence of interconnecting pores within monolithic Ti$_4$O$_7$ electrodes have accounted for most of the practical failures observed when using the material. They have also severely limited many other potential electrode applications. In general, electrolyte has the capacity to seep within the material and will creep above solution level to electrical connections (again, due to capillarity or "wicking") and all too easily jeopardise the integrity of the connection. The existence of porosity in anode materials is nothing new. A case in point is graphite, for many decades almost the only anode material used in chlor-alkali manufacture. At an empirical level, there were endless trials carried out, and patents filed, covering the impregnation by various means, of the graphite in order to eliminate porosity, in most cases using organic materials. In the main, these proved eminently successful. At the same time, there were numerous scientific studies and theories, some invoking physical effects, such as the build-up of high gas pressures within the pores, others involving chemical or electrochemical reactions leading to the degradation of the graphite.

Bearing in mind the graphite experience, it would seem logical to apply the same techniques to porous monolithic Ti_4O_7. Indeed it is practically straight forward to do so, both allowing the infillant to seep in by itself, or to use assisted filling by vacuum impregnation, applied external over-pressure or both. A wide range of organics have been tried; with particular attention focused on a carbon tetrachloride-based product, called Dichtol, because it was considered safer to handle than alcohol based types of product. It was also claimed by the makers that variants of the Dichtol product would withstand temperatures of 300° to 400°C without significant decomposition. This raised the prospect of infilling pores and then being able to coat the outer surface with a noble metal/oxide type coating with all applied material staying on the outside where it is required. Dichtol has remarkable penetrating properties and is widely used to infill hairline cracks in castings, for example.

Whilst pore in-filling does not adversely affect the electrochemical behaviour of graphite, the accumulated experimental evidence with monolithic Ti_4O_7 is to the contrary, with anode behaviour being badly affected. It took a long time, and there was some reluctance, to accept the reality of the findings. An example of such early trials, related to testing of uncoated and coated monolithic Ti_4O_7 electrodes in seawater at a coastal research centre (Brixham). The method of electrode construction is depicted in Fig. 2-14, consisting of porous strips, the connection end being potted with a standard CP potting resin. In an attempt to avoid seepage of electrolyte to the terminal connection, the part of the Ti_4O_7 adjacent to potting was repeatedly coated with Dichtol resin. In a few instances, rather than use of a potting resin, electrical connection was via titanium metal, bolted to titanium strip electrical connectors. Without going into too much detail of this rather large trial, electrode parts to which Dichtol had been applied, all failed unexpectedly rapidly, whereas those few electrodes without infillant did not deteriorate in the same way.

Such reluctance was there to accept that infillant could of itself be detrimental in preparing electrode assemblies, that several years later, for commercial products, the recommended technique for electrode connections was still to infill pores, then nickel electroplate, thereby bridging over the sealed pores, and then making solder connections. In most cases, such connector systems were not long lasting.

It is still not clear why organic infillants should cause problems in practice. There may be an analogy with the behaviour of aluminium exposed to hot carbon tetrachloride based cleaning and degreasing fluids. With non stabilised solvent, aluminium corrodes but not if the solvent has been stabilised. It is accepted that this is due to hydrolysis of the solvent, to form hydrochloric acid. Whether somewhat similar hydrolysis products are formed from the Dichtol which attack the substrate, is open to speculation. In this case, it might be that, in such a corrosion attack, bulky corrosion products could form, exerting internal pressures on the porous Ti_4O_7 and causing it to disintegrate?

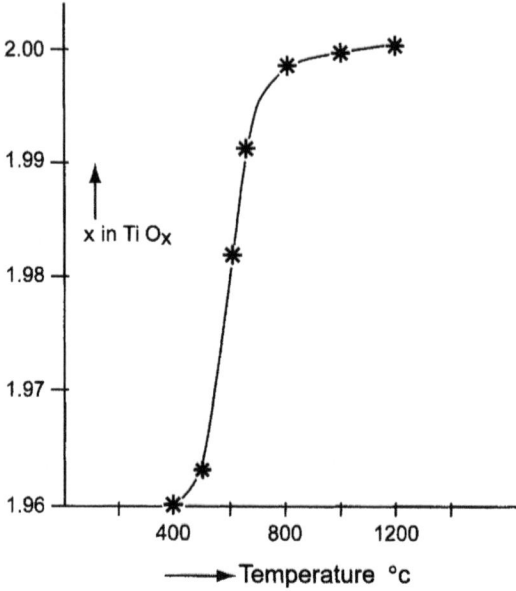

Fig. 2-15. Reoxidation of monolithic TiO$_x$ ceramic on heating 2 hours in air at increasing temperature.

Whereas difficulties have been encountered with use of organic infillents, and especially with chlorinated ones, the same has not been found for inorganic filling materials. The method of application is the same as used in much coated electrode production, namely so-called paint/thermal decomposition. Thus, for example tantalum pentachloride is dissolved in pentanol to a composition of 50 g/l metal content. Such solutions readily penetrate the porous Ti$_4$O$_7$. With solvent evaporated, the material is taken up to the decomposition temperature of the tantalum compound, which may be between 400° and 500°C. On conversion to Ta$_2$O$_5$, there is a volume contraction, so that even if pore filling could be accomplished in one so-called paint/stove cycle, then there would still be internal spaces left, so the process must be repeated several times. Practically it is unwise to saturate the porous Ti$_4$O$_7$ with paint, not only because it takes a long time for the solvent to diffuse out or evaporate, but there is a risk of overheating that completely destroys the product. Rather, the paint needs to be applied sparingly, and it may take 10 to 15 coat/stove cycles to ensure the product is substantially pore filled. There are, of course, alternatives to Ta$_2$O$_5$, commonly TiO$_2$. During such pore filling there will inevitably be some build up of material on external surfaces. Thus, if it is required subsequently to form on the outside surface, a coating of noble metal oxide or one containing mixed metal oxides, then specimens need an initial vapour blast treatment to expose the basic Ti$_4$O$_7$ structure.

A modification of the above technique is to infill pores with the same electrocatalyst as it is proposed to apply when pore filling is complete. This overcomes the necessity for an intermediate vapour blast treatment, and such filling also contributes to the electrical conductivity of the ceramic.

2.3 PLATING/COATING OF Ti$_4$O$_7$ WITH METALS AND/OR METAL OXIDES

Much of what is written below is to some extent amplification of information earlier noted.

With titanium as with most other metals, it is usual to pretreat surfaces prior to electrodeposition in order to improve coating adhesion. For titanium metal there are now a range of chemical etchants, but none so far has been found for monolithic Ti$_4$O$_7$, basically because of its excellent corrosion resistance. In part a pre-etching is not needed because the material is already so porous. A relatively simple pretreatment to apply if required is straightforward vapour blasting.

Platinum might seem an odd choice for initial electrodeposition trials, but work using this metal was carried out by a group of workers developing noble metal/oxide coated titanium electrodes. Using a sodium hexahydroxyplatinate plating bath operating at elevated temperature, no difficulty was encountered in securing adherent deposits similar to those which have in the past been electrodeposited onto titanium metal. Later, as expedient, copper, nickel, lead and lead dioxide coatings have been routinely applied. There seemed merit in particular in applying PbO$_2$ coatings to monolithic Ti$_4$O$_7$, because the coefficients of thermal expansion of the two materials are very similar, and thus might be expected to reduce the danger of the coating cracking.

The difficulty of applying noble metal/oxide containing mixed metal oxide by a paint/thermal decomposition rate has already been touched upon in connection with pore infilling. No easy way has yet been found of depositing electrocatalyst selectively on the outer surface without first filling pores. Techniques have included the use of very viscous, high concentration paints

A recently used technique applied to titanium metal may also be applicable to the coating of monolithic Ti$_4$O$_7$. Designed to improve the durability of electrodes in particularly corrosive conditions, the first step is to apply a platinum electroplated coating, preferably from a molten salt, and then to overcoat the platinum with conventional mixed metal oxide coating by repeated painting-stoving cycles.

Before concluding the subject of coating by paint/stove porous, an aspect to bear in mind is the thermal decomposition temperature of such paints. Some paints, to be preferred, can be satisfactorily decomposed at 400-450°, whereas others might benefit by raising temperature to 500°or higher. From the graphs in Figures 2-15 and 2-16, it will be seen that Ti$_4$O$_7$ exhibits significant oxidation at the higher temperatures. In order to deposit noble metal, as opposed to oxide, coatings, paints can conveniently be reduced to form the metal, at significantly lower temperature if heated in hydrogen, Ti$_4$O$_7$, unlike titanium metal, being unaffected by hydrogen below its forming temperature.

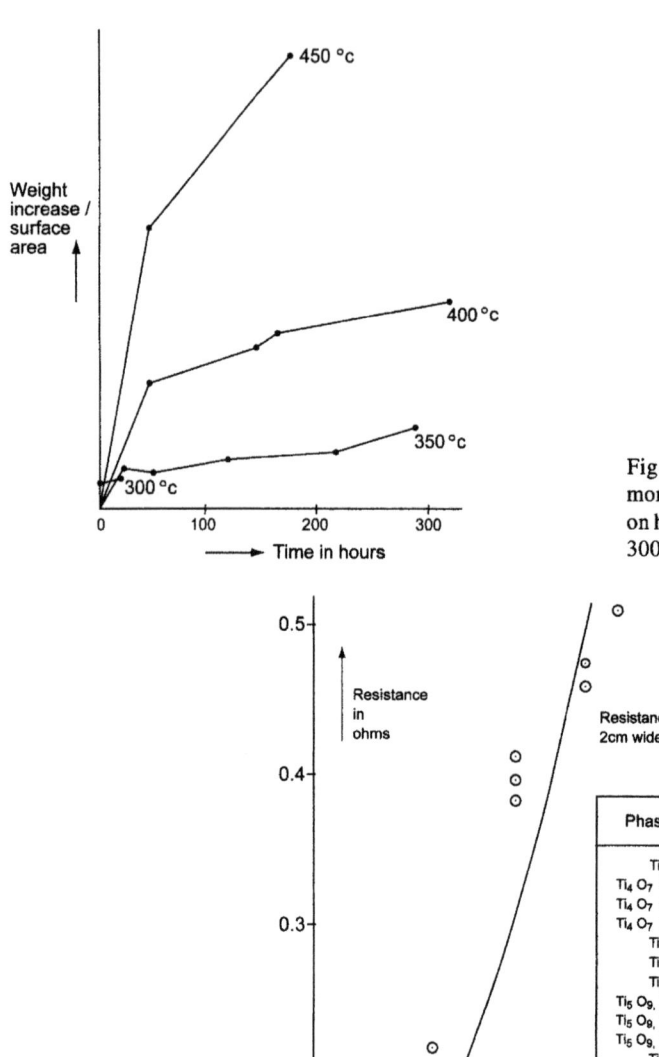

Fig. 2-16. Oxidation of monolithic Ti₄O₇ ceramic on heating in air between 300°C and 450°C.

Resistance measured on 6cm lenghts, 2cm wide x 4mm thick

Phases by XRD	Resistance ohms
$Ti_4 O_7$	0.038
$Ti_4 O_7$ 67%, $Ti_5 O_9$	0.046
$Ti_4 O_7$ 67%, $Ti_5 O_9$	0.057
$Ti_4 O_7$ 67%, $Ti_5 O_9$	0.053
$Ti_5 O_9$	0.11
$Ti_5 O_9$	0,071
$Ti_5 O_9$	0.108
$Ti_5 O_9$, $Ti_6 O_{11}$ 66%	0.094
$Ti_5 O_9$, $Ti_6 O_{11}$ 76%	0.126
$Ti_5 O_9$, $Ti_6 O_{11}$ 82%	0.053
$Ti_6 O_{11}$	0.199
$Ti_6 O_{11}$	0.188
$Ti_6 O_{11}$	0.216
$Ti_6 O_{11}$	0.098
$Ti_7 O_{13}$ 66%, $Ti_8 O_{15}$	0.385
$Ti_7 O_{13}$ 66%, $Ti_8 O_{15}$	0.396
$Ti_7 O_{13}$ 66%, $Ti_8 O_{15}$	0.413
Ti_8 41%, Ti_9 41%, Ti_{10}	0.476
Ti_8 41%, Ti_9 41%, Ti_{10}	0.46
Ti_8 41%, Ti_9 41%, Ti_{10}	0.374
$Ti_9 O_{17}$	0.512
$Ti_{11} O_{21}$	0.476

Fig. 2-17. Relative electrical resistance of a range of monolithic TiO_x ceramics of varying composition and from different sources.

2.4   THE ELECTRICAL RESISTANCE OF THE SUBOXIDE

The literature value for the electrical resistivity of Ti_4O_7 is circa $4000 \times 10^{-6}\ \Omega$ cm·
This can be contrasted with that of other titanium oxides (Figs. 2-17 and
2-18), and also with the resistance of other materials, (see Tables 2-7 and 2-8).
Graphite is only slightly more conductive than Ti_4O_7.

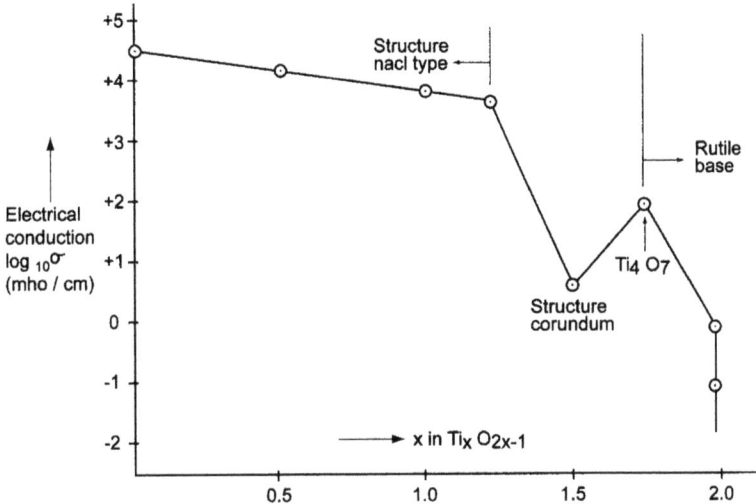

Fig. 2-18. Electrical conductivity of monolithic titanium oxides of
compositions between titanium and TiO_2.

TABLE 2-7   Properties of a Range of Oxides

Phase	Melting Point °C	Thermal Conductivity cal/cm/sec °C	Electrical resistivity Ω cm	Conductivity Type	Specific heat Ca/mol/°	Colour	Thermal Expansion Coefficient	Band gap volts eV
TiO	1820		1.7×10^{-4}	Metallic	10.6	bronze	9×10^{-6}	
Ti_2O_3	1930		0.5	semi <200°C metallic >200°C		blue/bla ck		0.06
Ti_3O_5	1850		6×10^4			brown/ black		
Ti_4O_7	1750	0.05	6×10^{-3}	metallic		black	8×10^{-6}	
TiO_2	1855	0.023	10^9	n-type semi	13.58	white	8×10^{-6}	3.04
ZrO_2	2677	0.05	2×10^4	semi	15.39	white	0.9×10^{-6}	
HfO_2	2840	0.004	$> 10^9$		17.39	white	6×10^{-6}	
VO	2050		10^{-2}	semi-metal	10.71	black		
V_2O_3	1970		10^{-3}	"	28.48	black		0.1
V_3O_5			1.0	"				
VO_2	1967			"	14.96	dark blue		0.8
V_2O_5	660		10^4	n type semi		orange		
NbO	1945		0.1	metallic	9.86	black		
NbO_2	1915		10^4	"	13.79	black		
Nb_2O_5	1491		$> 10^8$	"	31.59	white	2.6×10^{-6}	0.4
Ta_2O_5	1800		$> 10^9$	semi	29.2	white	1.6×10^{-6}	3.58

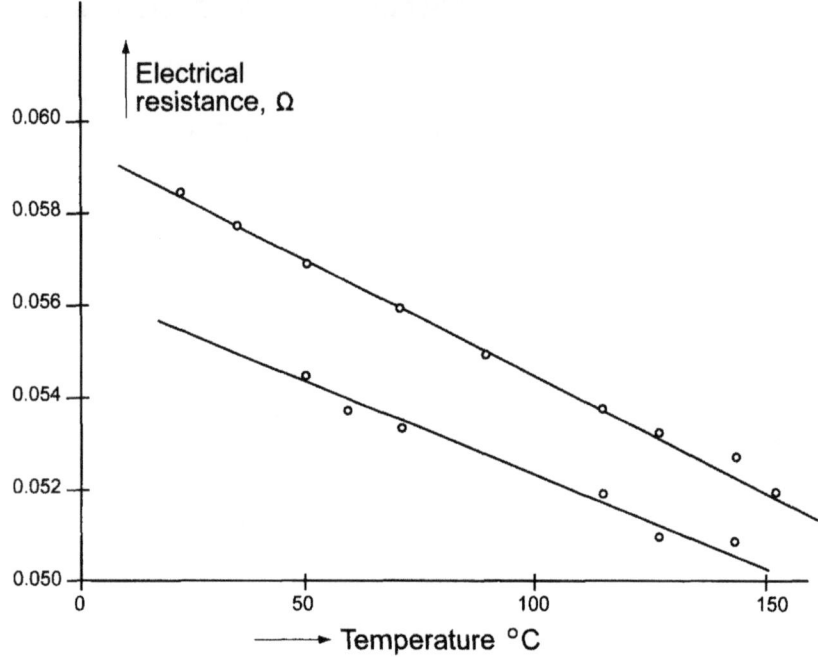

Electrical reistance measured between knife edges, for Ti$_4$ O$_7$ bar of approximate dimensions 4.5 mm thick by 14.5 mm.

Resistivity at   20°C   4689 × 10^{-6} Ω cm
                  150°C   4698 × 10^{-6} Ω cm

Fig. 2-19. Measurement of electrical resistance of monolithic Ti$_4$O$_7$ ceramic between 20°C and 150°C.

That monolithic Ti$_4$O$_7$ is indeed electrically conducting can be readily demonstrated by putting a sample in series with a battery and light bulb and showing that the bulb lights up! Other slightly less crude methods include resistance measurement using a multimeter. Use of such meters, which requires application of test probes to the sample, may give a reading which reflects surface resistance rather than bulk electrical resistivity.

One practical method of measurement of monolithic electrical resistivity involves passing a constant current down a bar of material, and then measuring the voltage drop between fixed points, (see Fig. 2-19)

TABLE 2-8

Properties of Various Materials

Material	Density g/cc	Electrical Resistivity Ω/cm	Thermal Conductivity at RT $WM^{-1}\,{}^{\circ}K^{-1}$	Coeff. of Expansion $10^{-6}/{}^{\circ}K$	Hardness $HV_{0.1}$
Alumina 99%	3.65/3.93		33	5.9	1900
Spinel MgAl$_2$O$_2$	2.8/3.2		~ 15	5.6	1500
TiO$_2$	3.5/4.0		2.5/4	5.7	
ZrO$_2$ CaO, MgO stab	5.0-5.8		1.7-2.0	8.9	
ZrO$_2$ Y$_2$O$_3$ stab	5.2-5.9		1.3	7-8	
Sialons (sintered Si$_3$N$_4$)	3.0			1.5-1.7	
Si$_3$O$_4$ hot pressed	3.1-3.2		7	1.5	1600-1800
SiC sintered	3.0-3.2		30-100	2.8	2500
SiC hot pressed	3.0-3.2		90-110	2.8	2400-2800
B$_4$C	2.3-2.5	10^6	30	3.3	3200
TiC	4.9	70	17		
BN hot pressed	1.9-2.1		20-30	0.1	soft
WC	15.5	20	60-80	4.7	1300-1600
BeO	2.8	$> 10^{14}$	300	5-7	1106-1300
Copper	8.9	1.7	398	16.6	
Aluminium	2.7	2.7	237	25	
Titanium	4.5	42	20	8.5	
Graphite	1.6-1.9	1400	5.9	3	soft
Ti$_4$O$_7$	2.5-4.0	5000			
TiB$_2$	4.5	30	26		
NbC	7.8	50	14		
TaC	1.9	70			

In another short series of experiments, see Table 2-9, comparison was made of the electrical resistivity of what became the standard monolithic T_4O_7, with material made from high purity starting TiO_2 powder. Evidently the starting material with the higher purity results in significantly increased electrical conductivity.

TABLE 2-9

Influence of TiO$_2$ Purity on Electrical Resistivity

Comparison was made between the relative electrical resistance of Ti$_4$O$_7$ made from standard ceramic grade titanium oxide powder and material supplied of much higher purity called TIL TiO$_2$ high purity.
Small compacts were made from the two sources of raw material, reduced in hydrogen at 1236°C and then examined by XRD for phase composition, and for electrical resistance at ambient temperature using a commercial 4 point probe meter. Comparative results were as follows:

Ti$_4$O$_7$ (high purity)	0.002 Ω
Ti$_4$O$_7$ (commercial standard)	0.014 Ω
Ti$_5$O$_9$ (high purity)	0.012 Ω

The evidence points to Ti$_4$O$_7$ ceramic made from a source of high purity starting powder resulting in significantly lower electrical resistivity compared with material made from high grade commercial ceramic grade TiO$_2$.

Nominal analysis of high purity TiO$_2$

Element	Maximum Content ppm	Element	Max Content ppm
V	10	Ca	20
CR	10	Al	10
Fe	10	Si	50
Ni	10	P	20
Nb	20	Sn	50
Na	20	Sb	50
K	10	Pb	10
		S	100

2.5 CORROSION RESISTANCE AND ELECTROCHEMICAL PROPERTIES

Early assessments of the corrosion resistance of monolithic T$_4$O$_7$ were made in precisely those solutions in which titanium metal does not perform so well, including in alkali, acid, or simulated zinc-winning electrolyte with 165g/l sulphuric acid and 15 ppm Cl⁻ + 5 ppm F⁻. To make meaningful comparison between the various titanium oxides, it was necessary to adopt a common form. All materials were thus prepared to powder form and sieved through a given mesh size. The compounds Ti_2O and TiO were arc melted prior to breaking down; The higher oxides were made by hydrogen reduction from starting TiO_2 powder. Results were similar, whether in alkaline or sulphuric acid based electrolyte, (see Fig. 2-20), corrosion resistance increasing towards the TiO_2 boundary. If such information is compared with electrical resistivity of the various oxide compositions, (Fig. 2-21), it becomes clear why the composition, Ti_4O_7, was chosen as optimum. This composition corresponds to optimum conductivity at highest corrosion resistance.

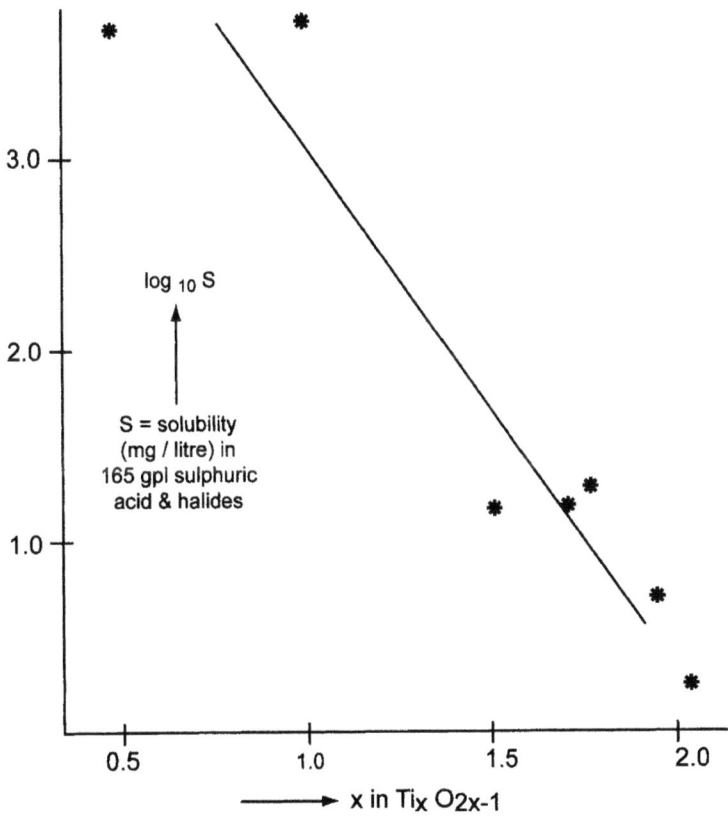

Fig. 2-20. Rate of corrosion in sulphuric acid-based electrolyte of various compositions of TiO_x ceramic (a similar trend is shown in strong alkali).

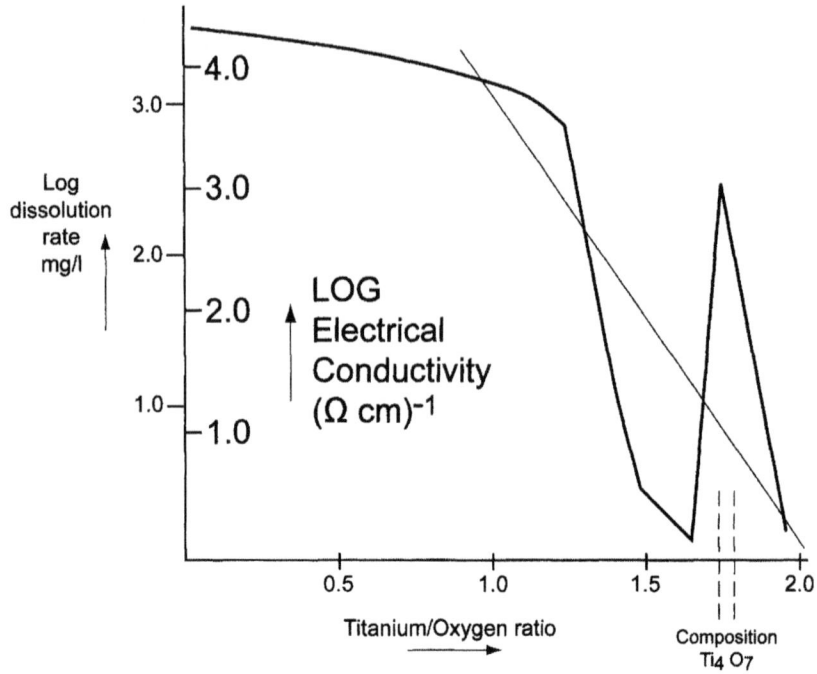

Fig. 2-21. Influence of titanium oxide composition on electrical conductivity and corrosion rate.

Thereafter, the behaviour of Ti_4O_7 in a range of media was systematically examined. Once material became available which was free from vitrifying agent, it was logical to think of hydrochloric acid, because many years previously, the preferred chemical etchant for titanium was three days immersion in concentrated acid at ambient temperature. During this period there was on average a removal of 0.0076mm per side, and formation of a thick sludge of titanium hydride. After 3 days immersion in concentrated hydrochloric acid, monolithic Ti_4O_7 showed no signs of attack and the solution remained colourless. Testing was continued over 2 years, with periodic change of solution, but still no corrosion was observed. Here, there, then, was an initial dramatic demonstration of the greater corrosion resistance of monolithic Ti_4O_7 as compared with titanium metal.

An acid causing enhanced corrosion of titanium is hydrofluoric acid, even in dilute concentration and at ambient temperature. Hydrofluoric acid containing pickles are widely used for titanium and its alloys. Therefore experiments were put in hand to compare monolithic Ti_4O_7 and titanium in various concentration of solutions containing hydrofluoric acid, (see Figs. 2-22, 2-23). The Ti_4O_7 was not immune from activation by hydrofluoric acid, but was at least 4 orders of magnitude more resistant than titanium metal, despite being so porous. How much better might corrosion resistance be for an homogenised, pore-free monolithic Ti_4O_7!

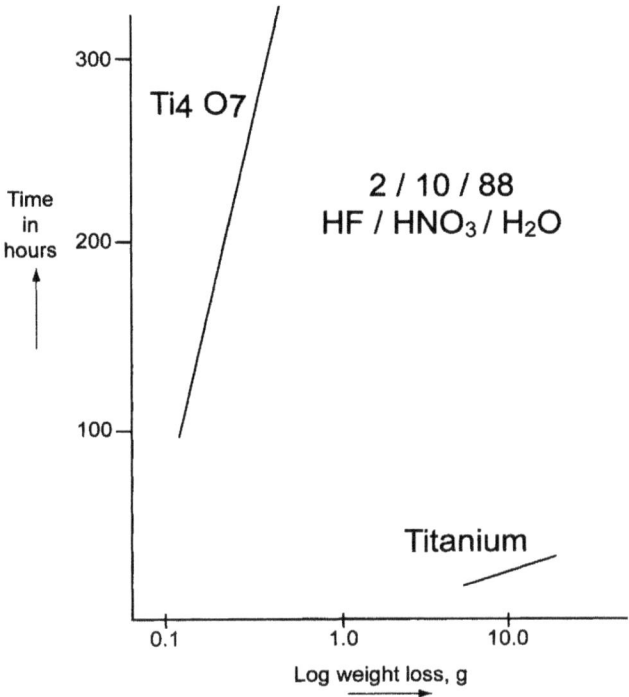

Fig. 2-22. Corrosion rate of titanium metal and monolithic Ti_4O_7 ceramic in a standard titanium metal pickling solution.

With the possibility of monolithic Ti_4O_7 in fuel cells containing phosphoric acid, appropriate testing was put in hand, (see Table 2-10), but with disappointing results.

The behaviour of monolithic Ti_4O_7 in hydrogen peroxide solution was strange, with the solutions sometimes becoming gelatinous.

As regards the electrochemical characteristics of the material, initial trials were designed to compare un-coated and platinum electroplated titanium with the sub-oxide in brine, (see Fig. 2-24). It became apparent that Ti_4O_7 possessed electrocatalytic behaviour in its own right, with a current carrying capacity intermediate between that for titanium, which almost immediately passivates, and noble metal coated titanium which is a superb conductor. There is evidence that freshly prepared Ti_4O_7 shows greater electrochemical activity than material stored in air over a period, which would seem to suggest that surface layers slowly revert to TiO_2.

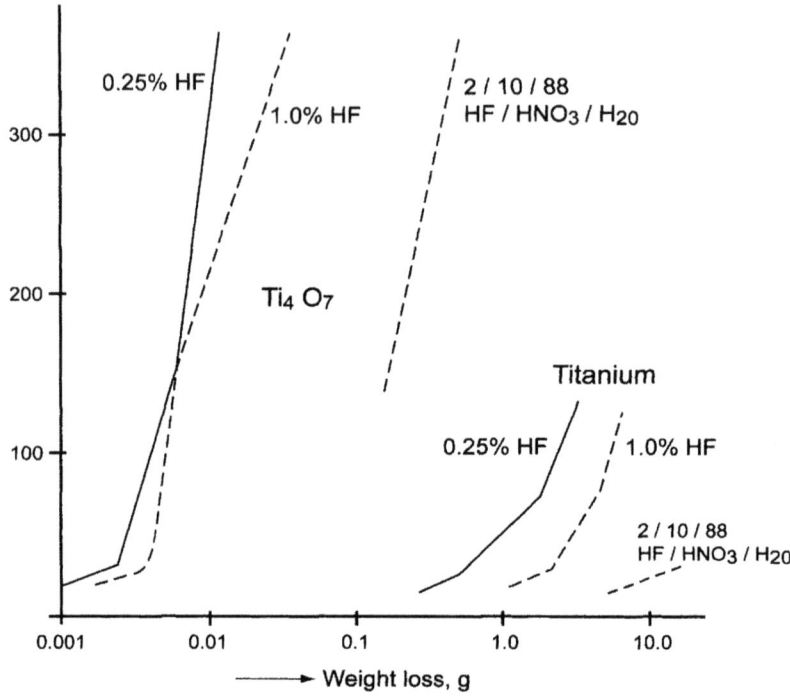

Sample size: Ti$_4$ O$_7$ (70mm x 30mm x 4mm)
Titanium (70mm x 30mm x 4mm)
Solution changed after 16, 33 and 75 hours

Fig. 2-23. Dissolution in hydrofluoric acid-containing solution of monolithic Ti$_4$O$_7$ ceramic and commercial purity titanium.

Monolithic Ti$_4$O$_7$ exhibits similar characteristics to titanium in brine as a cathode, but unlike titanium, does not form a superficial titanium hydride layer.

Some tests with monolithic Ti$_4$O$_7$ using periodic current reversal in brine showed an unusual behaviour, (Fig. 2-25). While the applied constant current switched polarity appropriately, the electrode voltage did not immediately follow suit.

The anodic overpotential of monolithic Ti$_4$O$_7$ is relatively high, as shown in Table 2-11. In this context the material can be considered broadly comparable to the recently disclosed diamond coatings formed from carbon in a discharge type reaction. Monolithic Ti$_4$O$_7$ requires care in making sound electrical connections, whereas the diamond coating performs best on a niobium substrate and is therefore likely to be costly.

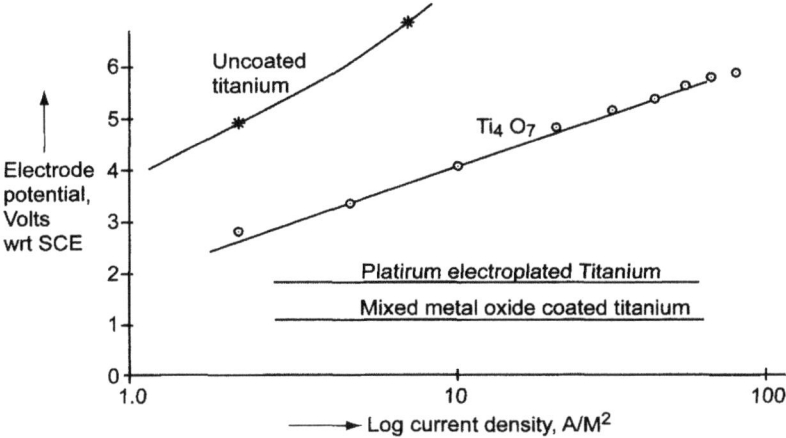

Fig. 2-24. Electrode potential versus log. current density for various materials, including monolithic Ti_4O_7 ceramic, in 10g/l brine at ambient temperature.

TABLE 2-10

Corrosion Resistance of Ti_4O_7

Orthophosphoric acid, strength ~ 90% at 190°C	Ti_4O_7 dissolves quite rapidly		
Caustic soda at 50°C	Initial sample wt., g	Wt. loss over 11 days, g	Ti in solution
30%	17.33 17.88	0.007 0.008	not detected
50%	16.76 17.95	0.021 0.024	1.2 mg
80%	16.50 18.44	0.087 0.070	1.8 mg

Thus Ti_4O_7 corrodes slowly in caustic soda, but increasingly fast with increasing concentration.

Included in Table 2-11, (see also Fig. 2-26), are some results for Ti_4O_7 nominally alloyed with different elements. Much caution is required in assessing such information because of variability that can affect production of starting material without detailed evaluation of material quality, including uniformity of alloying, the extent of porosity etc. but the indications are that certain additions can result in a lowering of overpotential. Alloying also affects thermal oxidation behaviour in air (see Fig. 2-27)

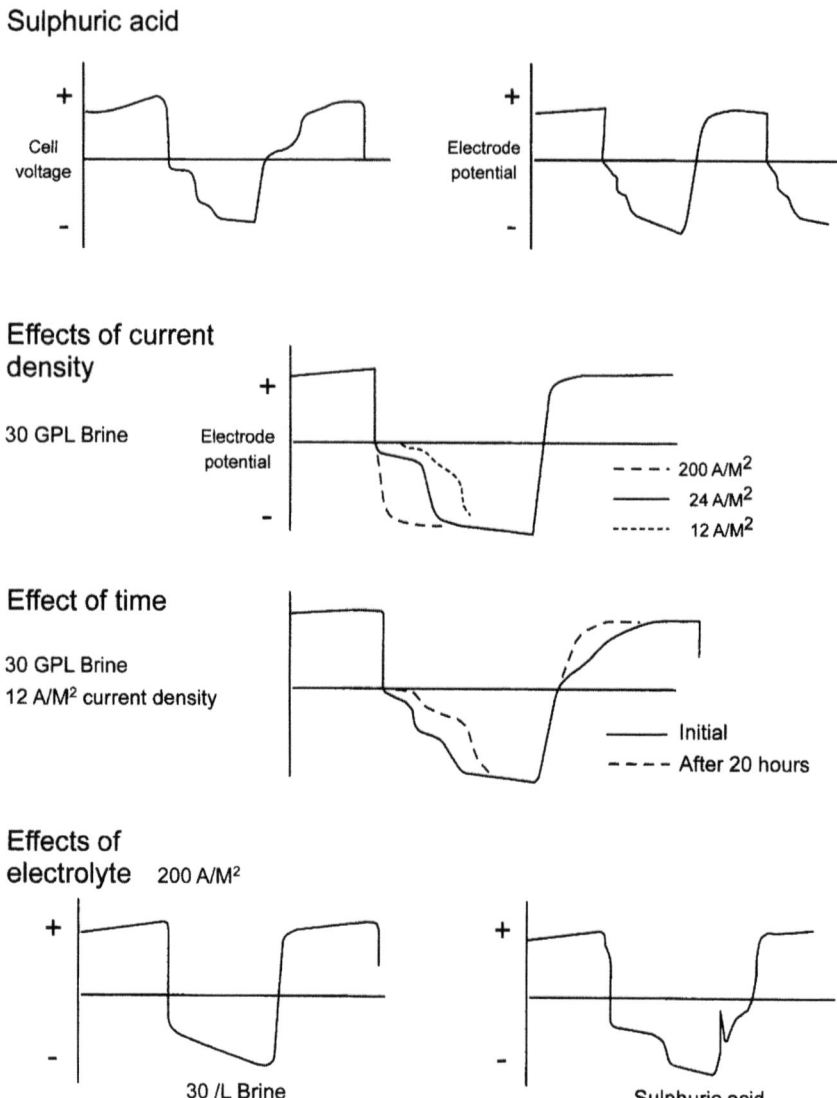

Constant current was applied to pairs of magneli phase titanium oxide electrodes, current was reversed in direction every 30 minutes. Corresponding changes in cell voltage and electrode potential were observed. The sulphuric acid electrlyte used was 165 GPL $H_2 SO_4$ + 115 PPM C1⁻ +5PPM F⁻

Fig. 2-25. Influence of periodic current reversal on the behaviour of monolithic Ti₄O₇ ceramic under various electrolytic conditions.

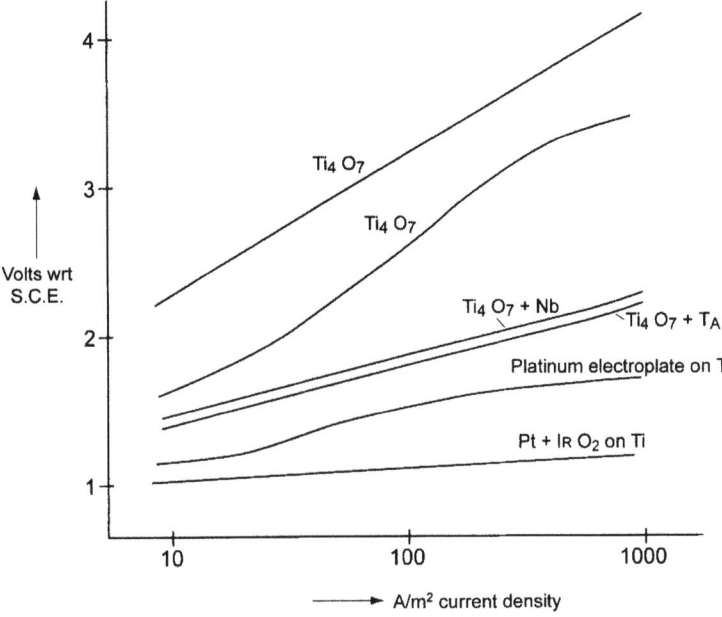

Fig. 2-26. Influence of alloying of monolithic Ti_4O_7 ceramic on electrode potential versus log. current density in 220g/l brine at 70°C.

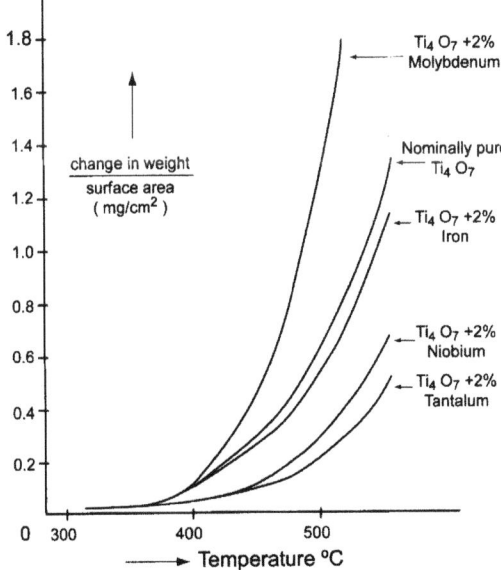

Fig. 2-27. Influence of alloying of monolithic Ti_4O_7 ceramic on thermal oxidation in air.

TABLE 2-11

Short Term Electrode Potential Measurements on a Range of Materials in Brine at Ambient and Elevated Temperature

Sample   Electrode Potentials, wrt SCE, in brine at 10 A/m² current density

	Ambient				70°C			
	200 g/l A	200 g/l C	30 g/l A	30 g/l C	200 g/l A	200 g/l C	30 g/l A	30 g/l C
Different samples, all nominally Ti_4O_7	1.62	- 1.44	2.11	- 1350	1.695	- 1.35	1.640	- 1.198
	2.19	-1.375	2.575	- 1.240	2.125	- 1.345	2.30	- 1.162
	2.23	- 1.45	2.12	- 1.345	2.20	- 1.371	2.25	- 1.221
			2.87	- 1.22			2.87	- 1.00
			2.25	- 1.55			2.150	- 1.334
Nominal Ti_4O_7 + 1% Ta	1.64	- 1.25	1.825	- 1.22	1.42	- 1.085	1.595	- 1.028
+1%Ta	1.475	- 1.14	1.73	- 1.13	1.37	- 1.135	1.590	- 1.104
+1%Nb	1.40	- 1.155	1.605	- 1.155	1.23	- 1.195	1.565	- 1.181
+1%Nb	1.83	- 1.281	1.99	- 1.225	1.42	- 1.080	1.500	- 1.039
+1%Fe	1.42	- 1.098	1.625	-1.085	1.24	- 1.039	1.330	- 1.037
+1%W	1.88	- 1.308	2.12	- 1.352	1.72	- 1.175	1.615	- 1.220
+1%W	1.83	- 1.28	2.00	- 1.300	1.60	- 1.252	2.02	- 1.234
+1%Mo	1.66	- 1.371	1.985	- 1.350	1.865	- 1.229	2.015	- 1.180
Magnetite	1.41		1.450		1.33		1.380	
Graphite			1.170	- 0.89			1.102	
Silicon iron			1.232	- 1.155			0.66	- 1.054
Pb+ I_xO_2 on Ti	1.71		1.090				1.040	
Platinum electroplate on Ti	1.125		1.156		1.08		1.070	
Stainless steel		- 1.249		- 1.265		- 1.152		- 1.119

2.6   ELECTRICAL CONNECTION TO MONOLITHIC Ti_4O_7

Initial thinking on how to make electrical connection to monolithic Ti_4O_7 treated the material as if it were metallic and therefore assumed that it should be electrically connected in the same way as for metals. If the material could be made non porous, this approach might well be fruitful.

Accepting that, for the present at least, commercially available material will continue to be porous, a logical approach, previously mentioned, might be to pore-fill, nickel electroplate and solder. In practice it has shown that such a procedure is less than satisfactory, as noted on page 28.

A better analogy in addressing the electrical connection problem, is with graphite and the techniques developed for making electrical connection to this material are essentially mechanical. Some of these are sketched in Figs. 2-28, 2-29, and though not exactly elegant in their conception, to date at least, these are proving both practical and sound.

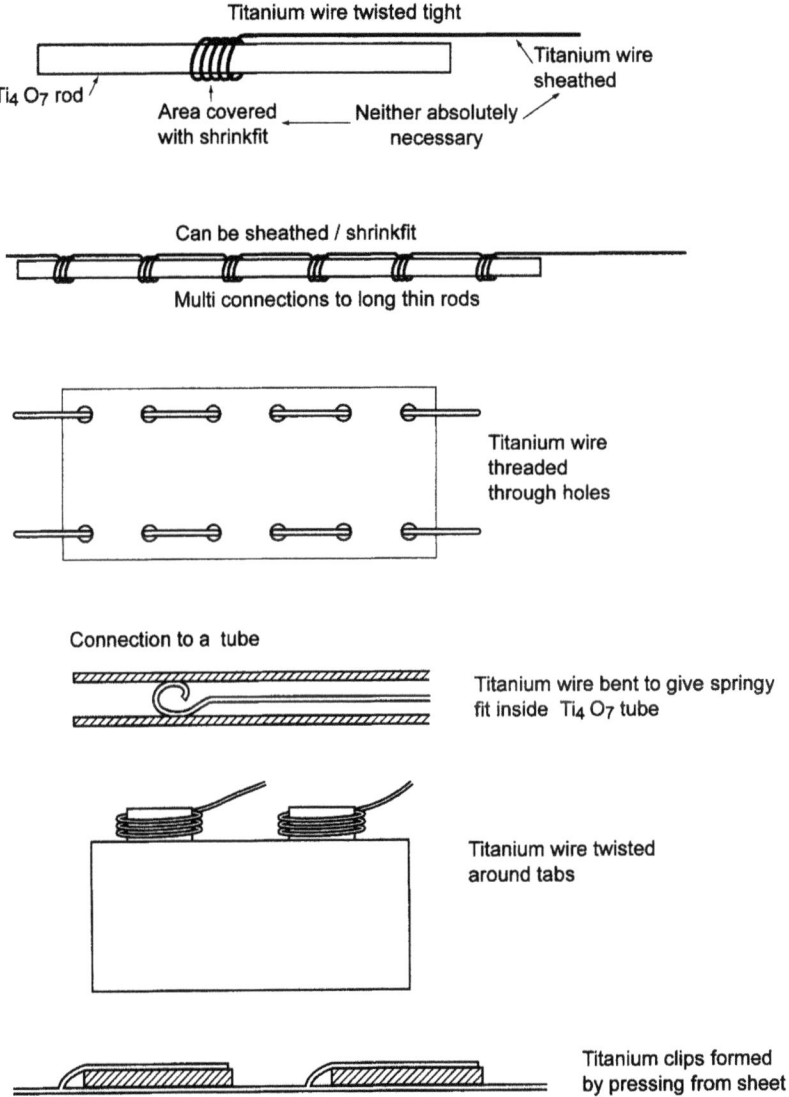

Fig. 2-28. Methods of making mechanical electrical connection to porous monolithic Ti_4O_7 ceramic electrodes.

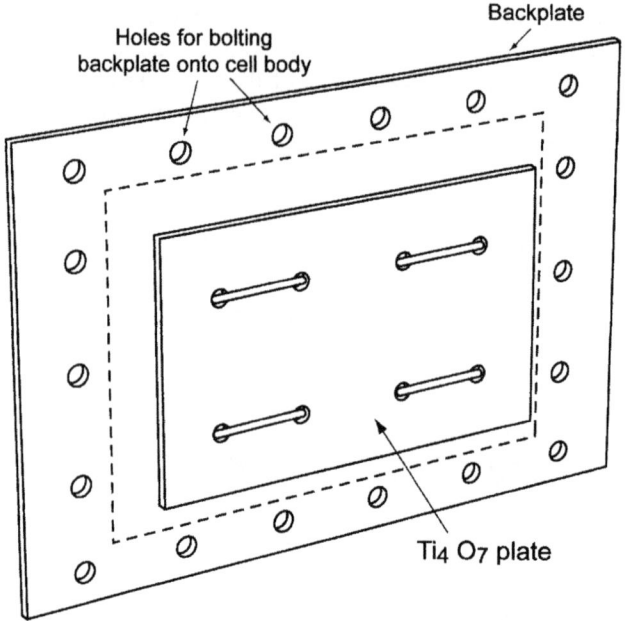

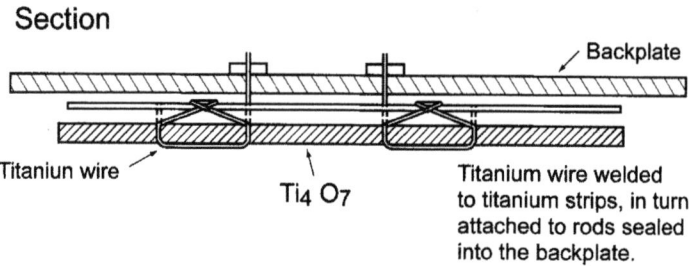

Fig. 2-29. Suggested method of incorporating porous monolithic Ti$_4$O$_7$ electrodes into electrochemical cells.

2.7 HANDLING/MACHINING

Hydrogen reduced TiO$_2$ is a brittle ceramic, but still much easier to handle than, for example, glass, no doubt in part because of the multiple internal defects. Such porous material is readily cut with a diamond saw, using water sparingly as a lubricant and coolant. Material can be drilled with diamond or silicon carbide drills, preferably at low speed and again using water sparingly. Monolithic Ti_4O_7 can safely be vaqua blast cleaned provided some background mechanical support is provided in the case of thin plate or tile. The material can be ground and polished for metallographic examination. The polished finish is an attractive black that might well find application in such fields as jewellery.

CHAPTER 3

3. Applications – I

When a new material becomes available, countless potential applications come to mind for its commercial exploitation. Manifestly, success or failure in a particular application depends upon the availability of the material in appropriate form. Many such early trials of "Ebonex" would now be well worth re-visiting, not only because the quality of the material has so greatly improved, but equally, because there is now a far better apppreciation of its properties and behaviour. There is little doubt that many new areas of application will be opened up, as non-porous forms become available.

3.1 ZINC ELECTROWINNING

As described in the introduction, there was a time when it seemed commercially viable to replace the massive lead-silver anodes used in electrowinning with a titanium-based product. One reason in particular, was the escalating price of silver, making it an attractive proposition to recover value latent in the silver content of the lead 1-2% silver anodes to set against the higher cost of titanium based anodes. As development of alternate anodes proceeded, the rising price of silver was exposed as fraudulent ....the notorious attempt by Nelson Bunker Hunt to corner the world silver market. With the subsequent collapse in the price of silver, (and decades later, the price has never remotely approached its peak at that time) so the financial case for titanium-based electrodes has been devastated. The issue is compounded because recently developed lead alloys with lower silver content are now additional contenders for this market.

Nevertheless, the development work at the time brought its own benefits as "spin-off". In the case of zinc, the electrolyte is a fairly strong sulphuric acid and usually also contains a few tens of ppm of hydrofluoric acid. This seemed an ideal opportunity to try to exploit monolithic Ti$_4$O$_7$ as a substrate because of its superior corrosion resistance to titanium. So began considerations as to how a ceramic could be incorporated into large area electrode construction.

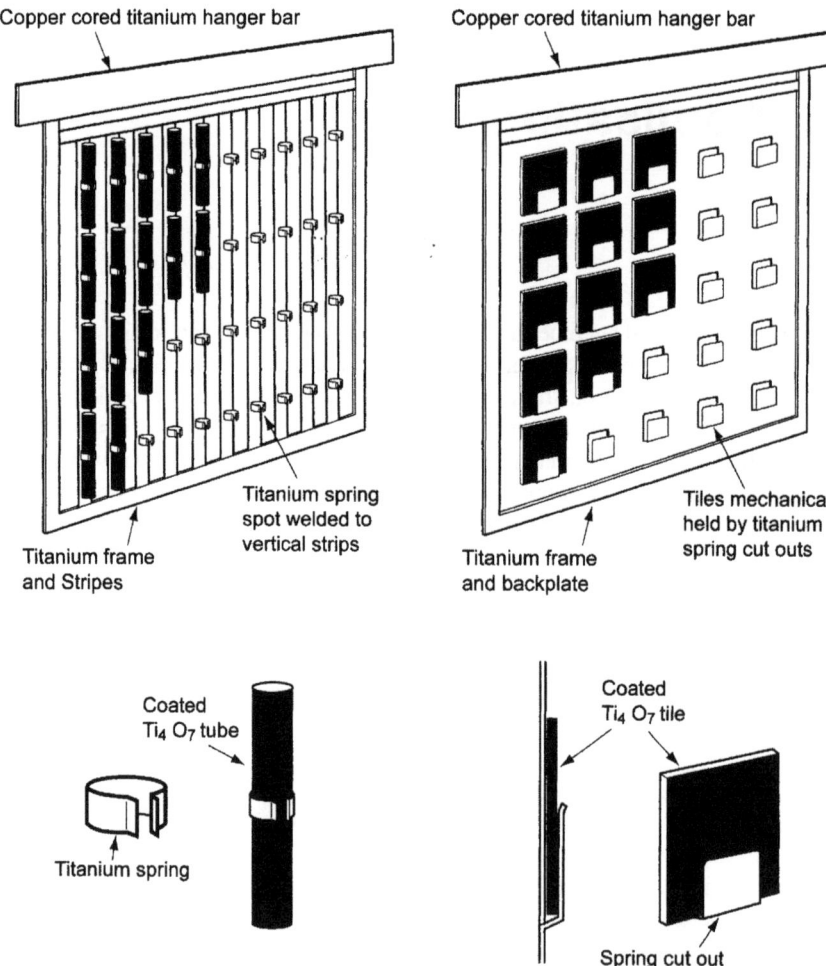

Fig. 3-1. Designs for use of noble metal/oxide coated monolithic Ti$_4$O$_7$ electrode material in large anode structures such as used in metal electrowinning.

Electrolytic testing of Ti_4O_7 coated with a Pt + IrO_2 electrocatalyst in simulated zinc winning liquor (165g/l sulphuric acid + 115 ppm Cl⁻ + 5ppm F⁻ at 35°C) revealed that the ceramic was indeed adequately durable and corrosion resistant, with testing up to a year before the project was abruptly curtailed. Possible designs for anode constructions are shown in Fig. 3-1. To reap the benefit of the lower overpotential characteristics of noble metal/oxide coatings, it was first necessary to remove the manganese ions, which would otherwise deposit onto the anodes as manganese dioxide, thereby obscuring the low overpotential noble metal/oxide surfaces. Indeed, it is by just such a deposition process that MnO_2 is industrially manufactured.

It was always recognised that even in the absence of manganese in solution, the IrO_2 anode surface would, in due course passivate, as the stoichiometric composition was reached, thereby raising the oxygen overpotential. The ceramic-based anode could be easily returned for recoating/reactivation and used again, as is normal practice in chlor-alkali plants. Such systems were pilot tested but did not reach plant-scale trials.

3.2 SWIMMING POOL ELECTROCHORINATORS

This is no trivial market, with the number of swimming pools constantly increasing and with a potential market requirement for tens of thousands of units. While each unit is required to produce quite small quantities of hypochlorite, compared with output from large commercial installations, nevertheless the technical problems associated with such small units have been formidable. For small-scale, domestically supervised units, it is unrealistic to expect owners to monitor, much less control, critical parameters such as water hardness or chloride concentration, or other water-quality related factors. The much reduced cost of electronic control circuitry offers the potential to automate the operation of such cells and make them more tolerant of abuse. The market demands trouble-free units which will operate over at least 4 to 5 or more seasons, and certainly not just one or two. As is so often the case, the goal-posts posed by competing technologies are not static, and chemical, chlorine-free, pool sterilisation technology has itself advanced.

A major problem with all electrochlorinators utilising natural waters is formation of hardness scale on cathodes. Such deposits can build to such an extent as to restrict electrolytic flow, and may even bridge to an adjacent anode and by electrical shorting cause damage to it. Going back almost 50 years, when platinum electroplated titanium was first being used in swimming pool electrochlorinators, principally in Australia, the practical answer to scale was periodic current reversal, say every hour or up to a day. The concept, which practically was very effective, leads, in the case of platinum, to enhanced dissolution and shortening of electrode life.

Another problem, less acute than cathodic scale formation, was cathodic corrosion of titanium which leads to thick film formation of titanium hydride. Occupying a greater volume than the titanium from which it is formed, thick hydride deposits cause warping of electrodes that can cause interference to electrolyte flow.

There followed a period when pools were in many cases, plastic lined and soft water was used. This avoided problems of scale formation due to water hardness, whether the latter was due to calcium or magnesium salts in the water supply, or species leaching out of the concrete, previously used in pool construction. Thus it seemed that the problem of scale formation had largely disappeared and the need for periodic current reversal with it. During this phase most platinum electroplate coatings were replaced by a mixed metal oxide (MMO). But still a large and increasing demand persisted for units capable of coping with cathode scale formation. Much ingenuity was shown in devising means for avoiding the necessity for current reversal, including mathematically designed fast flow systems, turbulence promoters, but although there have been literally hundreds of patents describing such ideas, nothing has so far emerged as more effective than current reversal. However there is no knowing what the future holds here. Electric kettles with "non-scaling" heater elements have been on sale for some years now, based on at least two different technological approaches. Meanwhile, the popular press and trade journals are flooded with advertisements and "advertorial" for scale inhibitors based on application of magnetic or pulsed electric fields. Opinion is divided between "experts" who state than such methods cannot, and indeed do not work, and those who have experience of them and find otherwise.

With the availability of monolithic Ti_4O_7, there seemed an immediate case for replacing titanium based electrodes in units operating with current reversal. Since Ti_4O_7 does not cathodically hydride like titanium metal, the major objection to current reversal disappeared, and would indeed the material need to be coated with a noble metal/oxide type electrocatalyst in order to function as an anode?

A first step was to demonstrate in the laboratory that monolithic Ti_4O_7 electrodes would work under a current reversal regime. Pairs of strip electrodes were polarised in brine, diluted to the levels normally used in swimming pools, with periodic current reversal. It was found from accumulated information, (see Fig. 3-2), that the material may be operable as an uncoated anode at current densities up to $100A/m^2$ .On the basis of such information, various types of experiment were initiated. In the one case, (see Fig. 3-3), two simple bipolar cells were constructed and fed with dilute brine to which had been added hardness salts. One cell was operated under dc condition, and the other with hourly current reversal. It was evident that current reversal was effective in dissolving cathodically deposited scale. In another trial, (see Fig. 3-4), a bipolar construction was assembled with plates set in resin. With plates of size 5cm × 5 cm, un-coated, and with hourly current reversal, the cell operated over many months. Hypochlorite was formed, though current efficiency was not high. An explanation for this is probably linked to the low current density employed. It appears to be that case on all known anode materials that chlorine evolution overpotential is less that than that for oxygen evolution. However the extent to which this is true, depends very much on the electrocatalyst (coating) in question, where some coatings are more "selective" in the extent to which they favour chlorine evolution than others. Table 2-11 illustrates this.

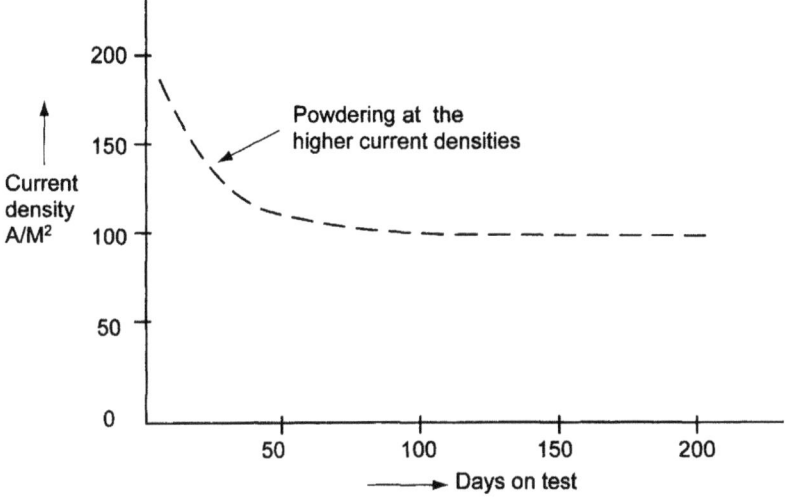

Fig. 3-2. Behaviour of monolithic Ti_4O_7 electrodes operated in 3% brine at ambient temperature and with half hourly current reversals.

TABLE 3-1

Comparative Impressed Current Anode Current Densities

Material	Application	Recommended Current Densities A/m²
		Current Densities A/m²
Magnetite, Fe_3O_4	Seawater	30-190
Graphite	Seawater	10
	Freshwater	$3 \rightarrow 4$
	Backfill	> 10, often $<< 10$
	Mud	
High silicon iron	Seawater	10
	Freshwater	10
Lead	Seawater	10
Platinum	Seawater	500 +
Plated Titanium	Backfill	$50 \rightarrow 100$
Mixed oxides	Seawater	500 +
	Backfill	100
Ti_4O_7	Seawater	10
Precious metal/MMO coated	Seawater	$100 \rightarrow 500$

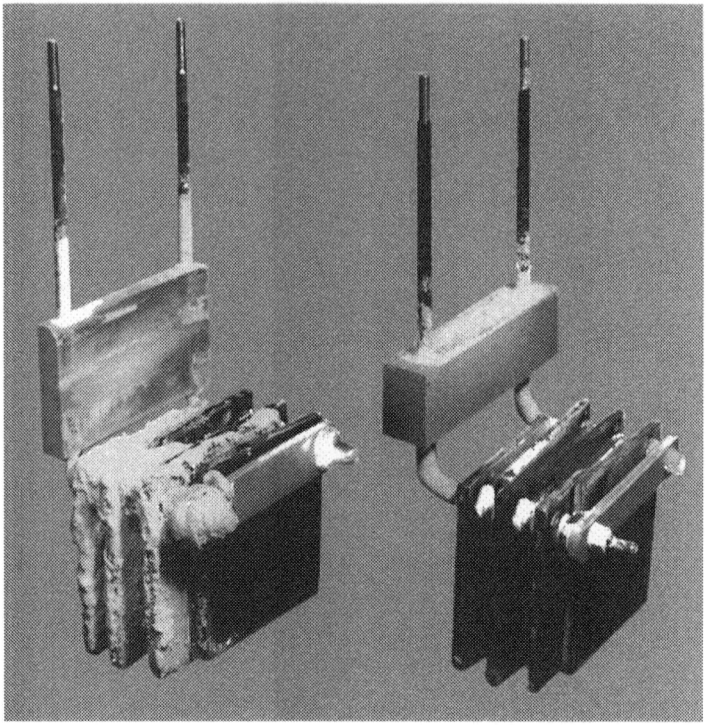

Fig. 3-3. Uncoated porous monolithic Ti$_4$O$_7$ electrode assemblies (not pairs) operated in 3% brine with added hardness salts. Left hand set: continuous current over 100 hours; right hand set: half hourly current reversal over 1500 hours.

As possible large-scale commercialisation of monolithic Ti$_4$O$_7$ in these electrochlorinators was being considered, the conclusion was reached that it would not be viable to do so, unless units could be operated at the similar current densities to those used with coated titanium electrodes, thus achieving comparable current efficiency. This necessitated the Ti$_4$O$_7$ being coated with established noble metal/oxide containing electrocatalyst. The problems associated with application of durable and efficient coatings have already been described. A further point should also be made here. As described above, Ti$_4$O$_7$ exhibits electrocatalytic behaviour in its own right. However it is nowhere near as good an electrocatalyst as the MMO and other precious metal/oxide coated surfaces. Because Ti$_4$O$_7$ remains a relatively high cost substrate (more expensive at present than the parent metal itself), there is an economic incentive to maximise the "productivity" of the material, that is to say the current flow per unit weight or unit volume. If application of a MMO or similar coating increases by ten-fold, the useable current for an anode of a given size, it may be economically advantageous to use such a coating. The truth of this applies, of course, not only to small electrochlorinators but to most devices using the material.

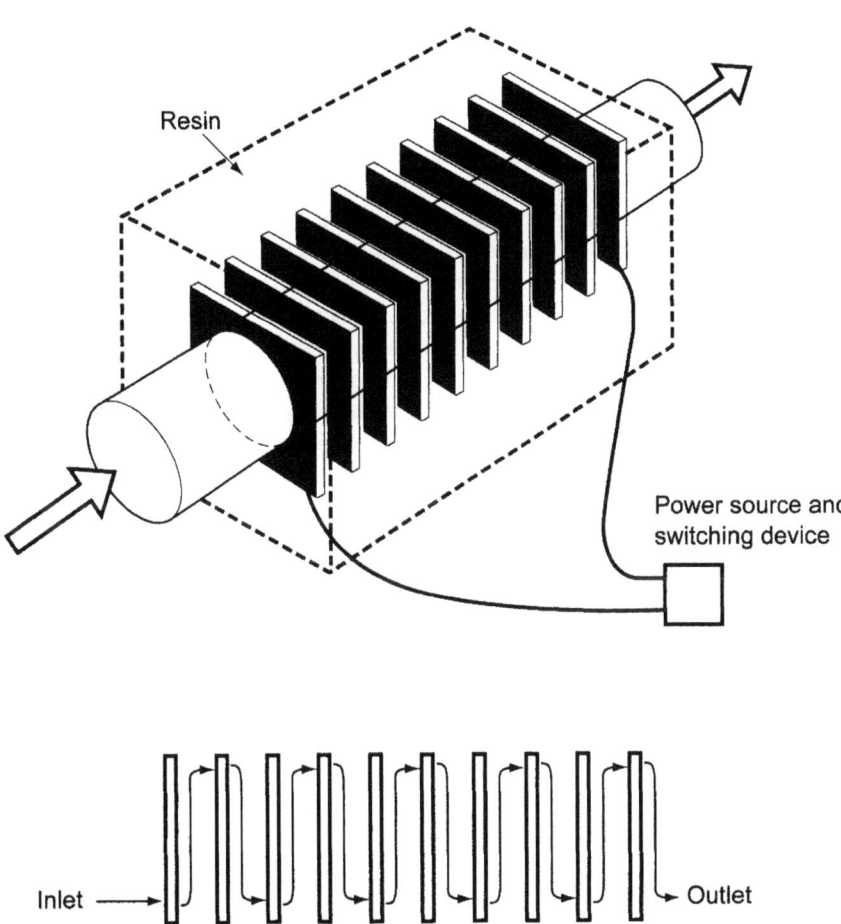

Fig. 3-4. Bipolar cell using 50mm x 50m uncoated porous Ti_4O_7 electrodes. Plates held in position with cast resin. Mechanical connection of titanium wire to monopoles. Cell operated at 1 amp. Half hourly current reversal in 3% brine plus hardness salts over many months.

However what ultimately proved the most troublesome aspect of the system, causing the overall concept to falter was the difficulty in making sound, long-life electrical connections to the monopoles . In the favoured cell design the monopoles had tabs outside electrolyte to which electrical connection was made by pore filling, nickel electrodeposition and soldering. How infuriating to consider, as described below, that simply winding titanium wire around the tabs, or threading wire through holes in these, would almost certainly have allowed the project to go forward.

In view of the above experience, there was an understandable reluctance for a long time to re-consider the use of monolithic Ti$_4$O$_7$ electrodes in reversing swimming pool electrochlorinators. At this time, a quantity of tiles became available that had been heavily coated both sides with what is known as a Beer II coating consisting of 30/70 wt/wt RuO$_2$/TiO$_2$. Not only were internal pores filled, but significant electrocatalyst had built up on the outer surfaces. Also conveniently available, were several empty plastic cell casings as used by an Australian manufacturer. (Monarch Industries). Using the coated Ti$_4$O$_7$ tiles, several cells was constructed, each incorporating several bipoles. In the first instance, electrical connection to the back of the monopoles was made by copper electrodeposition followed by soldering, (see Fig.s 3-5, 3-6, and 3-7). In later versions, a number of holes were drilled into the monopoles and electrical connection made simply by threading through titanium wire. Units were sent to South Africa and put into service in a near identical manner to that used for units incorporating titanium metal-based electrodes. Such units continue to function, those with soldered connections now in their 6th seasons, and those with titanium wire connection, and electrolytes at both the back as well as the front of monopoles, are in their 4th season. The experience may not necessarily prove the commercial viability of monolithic Ti$_4$O$_7$ in this application, but certainly testifies to the technical success of the two methods of electrical connection.

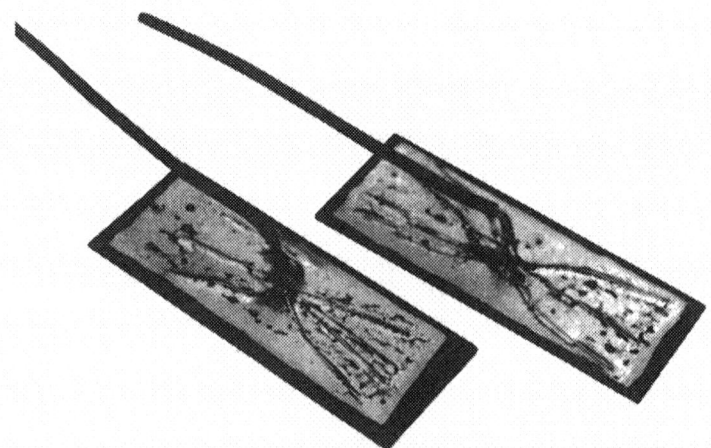

Fig. 3-5. Electrical connection to RuO$_2$.TiO$_2$ coated monolithic Ti$_4$O$_7$ plates by first copper plating followed by soft soldering.

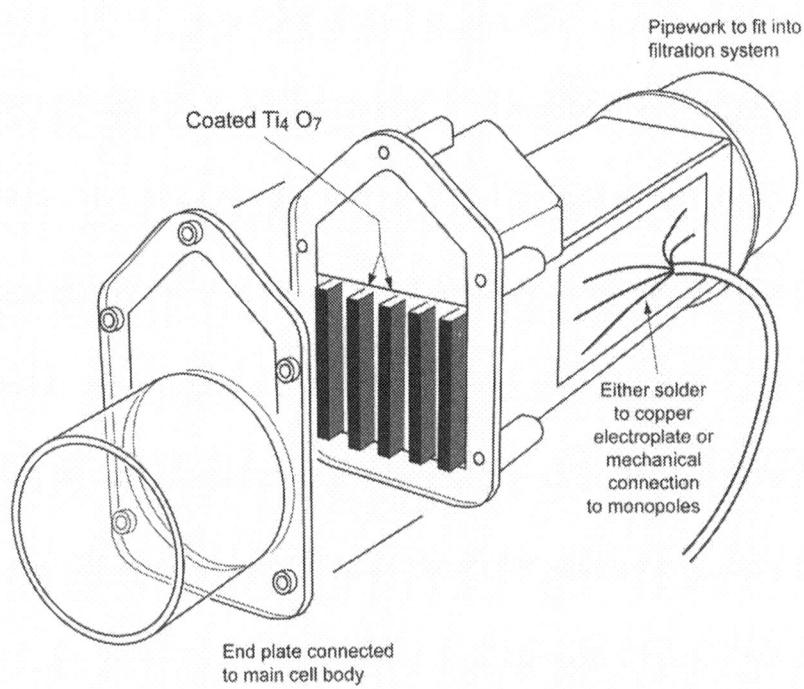

Fig. 3-6. $RuO_2.TiO_2$ coated monolithic Ti_4O_7 electrodes positioned in a commercial swimming pool electrochlorination cell body.

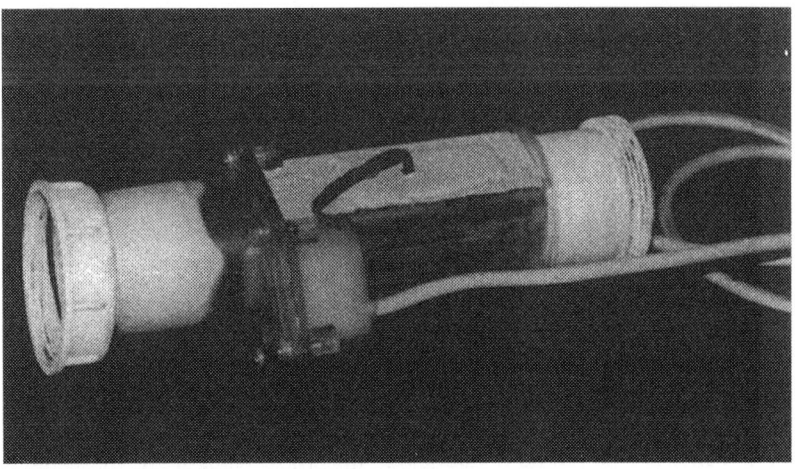

Fig. 3-7. Commercial swimming pool electrochlorinator cell body fitted with RuO_2TiO_2 coated monolithic Ti_4O_7 plates in bipolar mode. (cell casing courtesy of Monarch Industries, Australia).

An early problem with large scale electrochlorinators passing seawater or made up brine, was not just scale but also hydriding of the titanium cathode and subsequent electrode destruction caused by this. In one design of cell, massive block titanium electrodes were introduced to minimise the destruction problem. Monolithic Ti$_4$O$_7$ does not hydride (see Table 3-2) (Recall that the material is mostly made by hydrogen reduction) and should not therefore destruct like titanium. However, this attribute cannot be harnessed until non-porous material becomes commercially available, otherwise hydrogen would rapidly diffuse through such pores, reaching the coating on the anode side. As so often happens in an instance like this, the industry using titanium metal electrodes has sought to overcome hydride-induced deterioration problems in filter press-type bipolar designs by shifting to multi-monopolar designs where a cathode of some material other than titanium is employed to avoid hydriding problems.

TABLE 3-2

Hydrogen uptake of Various Materials After a Period of Cathodic Polarisation

Materials were made cathodic in 165 g/l H$_2$SO$_4$ + 115 ppm Cl$^-$ + 5 ppm F$^-$ at 35°C at a current density of 500 A/m^2. Analysis for total hydrogen after two weeks was as follows:–

Sample	ppm hydrogen	ppm hydrogen in non polarised blanks
Titanium	1880	5
Ti$_2$O	1430	50
TiO	73	5
Ti$_4$O$_7$	64	10

3.3 CATHODIC PROTECTION

Listed in Table 3-1 are some of the anode materials available to the cathodic protection engineer, with monolithic Ti$_4$O$_7$ for comparison. Why consider Ti$_4$O$_7$?. Whereas carbon/graphite will slowly be consumed under conditions of oxygen evolution, Ti$_4$O$_7$ should be more stable. The ceramic is more corrosion resistant than titanium in acids, eg the local acidity around anodes in mud or other deposit, and being close to the TiO_2 phase boundary does not suffer low voltage anodic breakdown corrosion.

First tests of monolithic Ti_4O_7 in seawater at a coastal seawater testing laboratory in the UK (Brixham) were far from encouraging. It became apparent after some considerable time that the reason for such poor initial results, as mentioned earlier stemmed from the use of organic infillants.

Despite the bad start , there seemed opportunity for trial of monolithic Ti_4O_7 in ground beds, and in particular the slim canister variety. The objective was to overcome some of the problem earlier associated with use of co-extruded platinum coated copper cored titanium wire. The ground bed is essentially a finely divided graphite to which electrical connection is made via a feeder wire. The feeder must be appropriately corrosion resistant, and while not of necessity an electrode in its own right, should preferably be so lest surrounding carbon is consumed or loses electrical contact with the feeder. For products using Ti_4O_7, rod was first made by extrusion into TiO_2 and later hydrogen reduced to Ti_4O_7 (size: 5mm OD $\times$ 760 mm long) Electrical connection was made to one end only by localised pore filling, nickel electroplating and soldering. Many thousands of slim canister ground beds (25 mm OD) are in operation, largely in connection with the cathodic protection of underground petroleum storage vessels.

For the future, it would seem more practical to make electrical connection to several parts of the rod by simply wrapping titanium wire around and sealing with heat shrinkable tubing. Not only would this give better electrical connection along the rod, but in case of fracture, electrical conduction would be maintained.

Using mechanical connection via titanium wire, a range of additional applications for monolithic Ti_4O_7 are possible, including its use in mud, tank bases etc.

3.3.1   Cathodic Protection of Rebars in Concrete

As background, the somewhat alarming discovery was made, in or around the 1970's, that much post World War II reinforced concrete was suffering from corrosion of the embedded mild steel rebars ("rebar" is a widely-used abbreviation for "reinforcing bar") mainly due to salt migration from roadways but also from atmospheric pollution. Short of sometimes extensive repairs, or even replacement (both of which required that much of the concrete mass be broken up), what could be done to stem further ongoing corrosion? The Federal Bureau of Highways in Washington gave an important lead, much against the wishes of the civil engineering lobby, with their verdict that impressed current cathodic protection offered the best chance of arresting further deterioration. The recommendation opened the floodgate to a range of methods for implementing cathodic protection (often referred to as "C.P"). One of the earliest and cheapest methods was to apply to the outside of the structure, a carbon loaded, electrically conducting paint which was then made anodic with respect to rebar. Such a technique is still used, but a consideration is that projected life time of the method, perhaps 5 to 10 years or more, is not as long as can be obtained using some of the alternative methods. The most widely used of the projected longer life methods is the use of electrocatalyst coated titanium chicken wire type mesh strip. Laid over the outer surface of the concrete, electrochemical connection to the base structure is made by overlaying the mesh with a cementitious coating, which also acts to keep the mesh physically in place and protect it from mechanical wear. Such repair mechanisms have projected life times of 20 years or more.

Against this background, it might not seem necessary to search for further alternative techniques, but in a situation where a range of patents protect many of these methods (though many of these will soon expire), questions of cost and meeting new requirements, many other techniques have been devised. Could monolithic Ti_4O_7 play a role in any of these techniques? It was recognised that use of carbon and graphite would mean slow consumption should the anode evolve oxygen, since under anodic conditions, these species break down electrochemically (to form carbon dioxide) and also mechanically. Would Ti_4O_7 in particulate form offer any advantages as a pigment loading in electrically conducting paints.?

With the assistance of a paint manufacturer, a quantity of proprietary paint as used in the CP of rebar in concrete was procured and analysed to determine the dispersion system. Rather surprisingly the pigment contained a small quantity of slate as well as graphite, presumably added to decrease costs. To ensure meaningful comparisons, all paints made up with Ti_4O_7 used, as far as possible, the same formulation.

A control was made up using graphite and found to exhibit similar electrical conductivity to that of the commercial product. Next a quantity of Ti_4O_7 powder was obtained via hydrogen reduction of TiO_2 starting material. The particle size proved far too coarse, so heavy milling was carried out to create a particle size distribution comparable to that of the graphite in the proprietary paint. Results obtained on testing the various paints, (see Table 3-3), suggest that paint made up using Ti_4O_7 is very significantly less electrically conducting than proprietary graphite loaded paint. As is so often the case in industrial development work, it was not possible to examine this issue in a way that would allow an unequivocal conclusion to be drawn. Thus heavy milling could result in heating of the powders, and thereby accelerate a re-oxidation of fine particles which in any case have a very high surface area:mass ratio. Paint made up with Ti_4O_7, in this way, while rejected at the time for use in the CP of rebars, might nevertheless find application as an antistatic coating, priming of plastics prior to electrodeposition and similar applications. In view of the close similarity of electrical resistively values for bulk graphite and Ti_4O_7, the disparity between the two types of pigmented paint seems all the more surprising. Compacted Ti_4O_7 particles, (see Fig. 2-9), exhibit electrical properties similar to those characteristic of monolithic Ti_4O_7. It might be that the action of crushing disrupts the insulating high oxidation state surface films, presumably of TiO_2, and these, deprived of adequate access to oxygen, acquire a degree of electrical conductivity characteristic of the suboxide.

TABLE 3-3

Ti_4O_7 Powder Versus Graphite in the Manufacture of Electrically Conducting Paints

A commercial electrically conducting paint was analysed. The PVC content (pigment by volume) was high at ~ 60% (c.f. ~ 20% for household paints). Because of the high pigment content, the resin content was low meaning a 'matt' as opposed to 'gloss' finish. 20% of the pigment content was 'slate' dust rather than carbon.
A similar paint was made up to that obtained commercially, adding carbon and slate dust.
A third paint was made up using Ti_4O_7 powder, particle size passing through a 100 m size mesh, using the same vehicle and PVC content as the commercial paint. Relative resistance measurements were:–

Commercial paint	2000 Ω	} similar within
Simulated commercial paint	1700 Ω	} experimental error
Ti_4O_7 based paint	5,000,000 Ω	i.e. × 2500 more resistive.

Preparations were made by a paint manufacturer. While Ti_4O_7 particle and grain size might still be variables for investigation, chemical activation of the surface layers of the Ti_4O_7, different resin/dispersion systems, the indications are that Ti_4O_7 will not easily be made into the basis for an electrically conducting paint.

As with so many proposed use of any new material, in the case of monolithic Ti_4O_7, it has to compete against established materials in seeking new applications. A further opportunity for possible use of monolithic Ti_4O_7 came when, in the UK, the Department of Transport became more interested in the possibilities of impressed current cathode protection of motorway bridges. Despite the considerable experience of this technology, already gained in other parts of the world, the British Ministry of Transport (as it was), were anxious to obtain first-hand experience under UK conditions. A comparative trial of several techniques was therefore set up at the Midlands Motorway Link at Gravelly Hill (Spaghetti Junction). Proposals were invited, with close questioning as to how proposed techniques would address any problems of local acidity generation around anodes, and the mode of release of electrochemically generated gas. The test location was conveniently close to the laboratory where the monolithic Ti$_4$O$_7$, was produced, and it was proposed to insert porous tubes at approx. 250mm intervals on all sides of a cross beam supporting the motorway, (see Fig. 3-8). It was felt that the electrochemical activity of uncoated Ti$_4$O$_7$ would be adequate for the purpose, with electrical connection between tubes using titanium wire. It was argued that normal rainfall would dilute any acidity forming on the Ti$_4$O$_7$/concrete interface, and of course allow ready release of any anodically generated gases. A further suggestion was that drainage from the roadway above, be allowed to trickle over the inset tubes to dilute any anodically generated acidity.

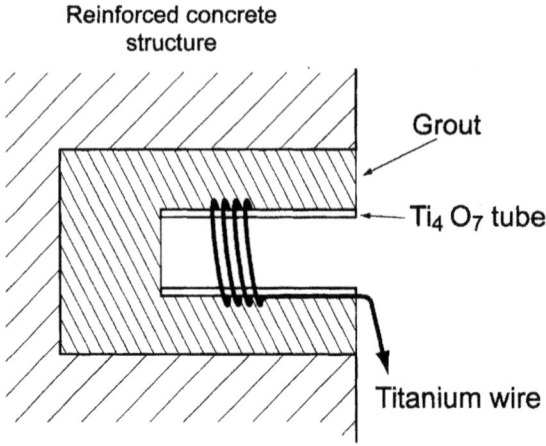

Fig. 3-8. Initial design, subsequently rejected, for inserting porous monolithic Ti$_4$O$_7$ tubes as point anodes in the cathodic protection of mild steel rebars in reinforced concrete.

This proposal was not accepted by the authorities. A principal objection was that the cover on the cross beams was variable and in some locations so thin, that drilling holes could expose the rebars. Furthermore, there was a fear of drilling so many holes, ~ 2000 per cross beam, that the structure would be weakened. In a modified proposal, which was accepted, monolithic Ti$_4$O$_7$ tiles, each 5cm × 5cm × ~ 3mm were cemented to the surface, at 250mm intervals, electrically connected by a 2cm wide titanium strip, (see Fig. 3-9). The tiles were uncoated and highly porous to aid dilution of anodically generated acidity (see Figs. 3-10 and 3-11). The current necessary for protection of a single cross beam was ~ 2 amps. Between concept and application there was stringent laboratory testing of all aspects of the installation, including repeated freeze/thaw trials of Ti$_4$O$_7$ tiles saturated with water. These showed no indication of physical break-up of the material due to the well-known expansion of ice on thawing. The method used is illustrated in a photograph (Fig. 3-12).

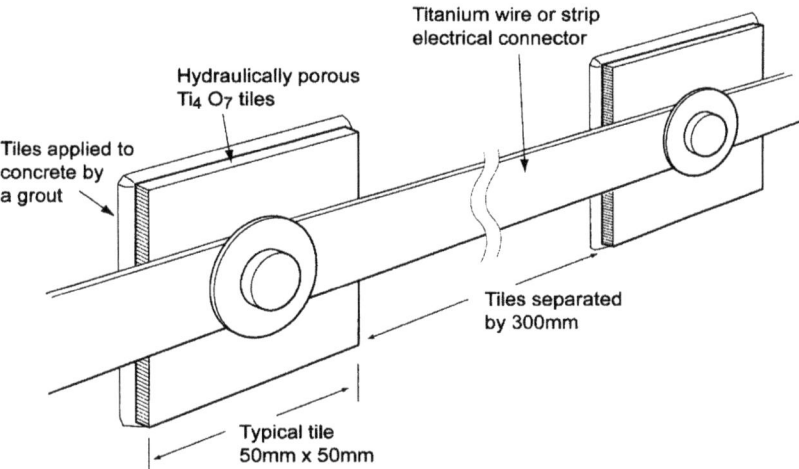

Fig. 3-9. An accepted design for attaching porous monolithic Ti_4O_5 tiles to reinforced concrete as part of a cathodic protection system to arrest rebar corrosion.

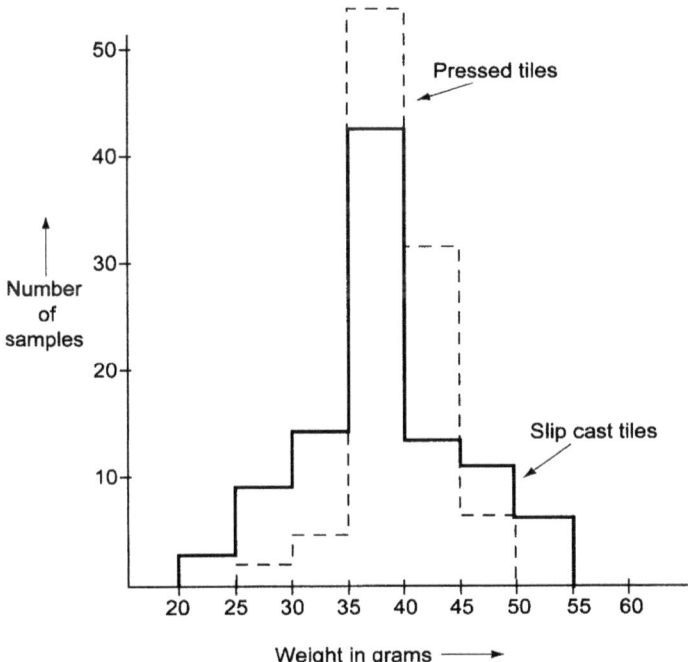

Fig. 3-10. Weight variation of nominal 50mm ×50mm monolithic Ti_4O_7 ceramic tiles first made in quantity for use in the cathodic protection of rebar in reinforced concrete.

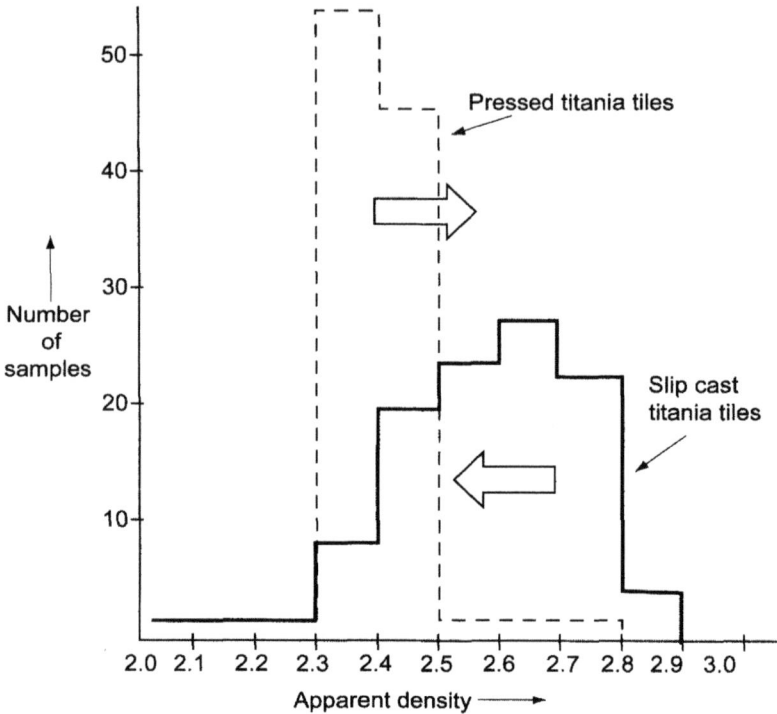

Large arrows indicate apparent density changes during hydrogen reduction to Ti4 O7.

Fig. 3-11. Density variation, and thus inferred porosity, for monolithic Ti$_4$O$_7$ electrodes first made in bulk from two manufacturers for use in the cathodic protection of rebars in reinforced concrete.

In this saga of the commercial exploitation of monolithic Ti$_4$O$_7$, the trial did not work as satisfactorily as anticipated. But the system was maintained under polarisation for over 5 years, during which there was no loss of adhesion of tiles nor any obvious physical damage. At best the system allowed passage of up to two-thirds the design current rating. The weakness lay in thinking that the cross beam would be substantially rain irrigated. In practice, the cross beam was largely sheltered by the roadway, with only the ends of the cross beam protruding on either side of the roadway receiving rain. Furthermore, the cross beam was totally exposed to the wind, which exerted a drying action. The problem was that not only did the draughts dry out the porous Ti$_4$O$_7$, but more significantly, the highly porous Ti$_4$O$_7$ drew moisture from the grout, thus greatly increasing its electrical resistivity. Where surface tiles were rain-irrigated, current passed selectively, and evidence accumulated of acid attack on the underside of each tile. This is readily observable as acid attack reveals and makes visible individual sand particles, which can be white, yellow or red, depending on the source of sand. In this case, the acid-attacked material showed pink coloration. (see Fig. 3-12).

Fig. 3-12. Cathodic protection applied to a cross beam supporting a motorway, using surface mounted, uncoated, porous monolithic Ti_4O_7 anodes.

In retrospect it may well be thought that all the problems associated with the above quite large scale and much publicised test should have been foreseen, but nevertheless valuable experience had been gained. The adverse public relations aspect of the project made it unlikely that further opportunities of this kind would be offered, but the concept based on the use of surface tiles in a situation where concrete is constantly exposed to water, for example tidal washed bases of bridge pillars in estuaries, would appear to be both technically viable as well as commercially attractive.

In more recent years, a technique for CP of rebars in concrete has come to the fore using so-called point anodes. Initially these were shunned by the industry as impractical for application to large surface areas. But specific needs arose in connection with application of CP to buildings and also to extend the throwing power and consequent protective action of CP to regions well below the surface, especially in connection with joints. The point anode became popular especially in Scandinavia. An early design of point anode was in effect a mini ground bed. A hole would be drilled into the structure and filled with carbon or carbon containing material, to which electrical connection was made by inserting a platinum electroplated titanium rod of a few mm diameter. With time, not only has the composition of the carbon electrocatalyst used in this manner been modified, but the platinised rod has been replaced by a mixed metal oxide coated titanium wire. More recently the carbon has been omitted altogether and the MMO coated titanium wire or tube grouted directly into the structure.

With the obvious success of the point anode for certain types of repair to reinforced concrete, and initially on a small scale, there was a resumption of the evaluation of Ti$_4$O$_7$ tubes set into concrete, mainly from the point of view of assessing different methods of making electrical contact. In the initial designs, such contact was merely a bent piece of titanium wire pushed securely down into the tube. As an adjunct, the tube interior was filled with a proprietary ground bed material, Loresco™ and tamped either end with silicone rubber as sealant. The view at the time was that such a method might make better electrical contact along the lengths of tubes varying from 5 to 10 cm in length. As the years of polarisation elapsed, and affecting first those tubes operating at the highest current density, the Loresco powder was observed to have been forced to the outer surface, pressured by the internal gas generation, small though it might have seemed when expressed on a daily basis. Also, the telltale pink colour showed there was acid attack occurring to the concrete immediately adjacent to the Ti$_4$O$_7$. Further installations omitted the Loresco, and put in its place a specific gas venting tube, at first a drinking straw and then polythene tubing. What started as a project to study means of making electrical connection to an anode, turned out to be one of studying gas release and local acid attack. Evidence occurred to show that venting of gas also released acid vapour, hence minimising internal acid attack. All methods of making electrical connection, including melted lead, were found practicable. Latterly, connection was simply made by twisting a piece of titanium wire around the centre of the tube and covering with a short length of shrinkfit, and this also is proving practically adequate. The monolithic Ti$_4$O$_7$ tested in concrete at one location has now been passing current satisfactorily for up to 6.5 years so far.

Given that uncoated monolithic Ti$_4$O$_7$ tube was adequately passing current after a number of years there seemed no reason why such items should not be used commercially in competition with the other available materials and techniques. Indeed now many tens of thousands of such point anodes are in service, and this must therefore rank as one of the more important uses of Ti$_4$O$_7$ to date.

CHAPTER 4

4. Applications – II

4.1 ELECTRODE BOILING

Early steam generating plant employed a principle known as electrode boiling. AC current is applied between two electrodes at a high frequency sufficiently high to prevent any Faradaic reaction from proceeding. Heat is generated by the IR voltage drop between electrodes. Usually 50Hz AC is used but some much higher frequency high current density systems have been developed for rapid sterilisation of foodstuffs. A particular attraction of electrode boiling over and above other methods is a safety factor. If the electrolyte boils dry, then no further current passes and hence over- heating is avoided. Electrode boiling has attraction for domestic appliances, and in one design, the electrodes did not face one another, but were inlaid side by side, on the floor of the kettle.

Not all materials are resistant to electrode boiling. Surprisingly the noble metal platinum succumbs to corrosion at 50Hz when used simply to boil water. Monolithic Ti_4O_7 behaves well under AC, possibly because the cathodic cycle is helpful in preventing any tendency to anodic passivation. In connection with electrode boiling, pairs of strip electrodes were used to boil water in a reflux apparatus (see Fig. 4-1). The testing was continued over many months without apparent deterioration of the electrodes.

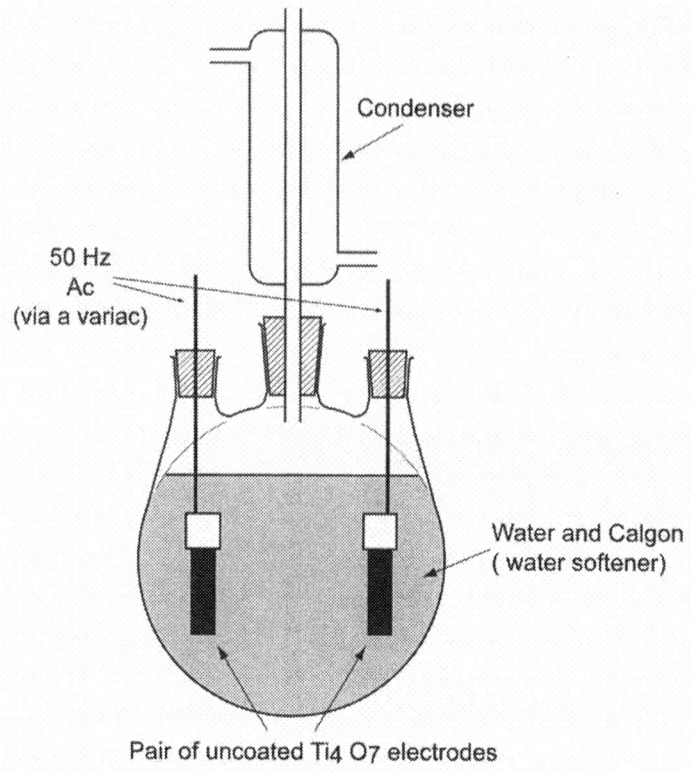

Fig. 4-1. Laboratory test setup for prolonged AC heating of water using uncoated monolithic Ti_4O_7 electrodes.

The successful experiment described above, does not necessarily imply that monolithic Ti_4O_7 is a contender for commercial application, unless it is acceptable to use only moderate current densities. Hence for possible application for instant water heating, as might be required in cloak rooms, trains etc, where 3 phase may be necessary to supply sufficiently rapid heating, (higher energy densities can be supplied by three-phase than using single phase supplies, and three-phase is available on most railway coaches). Under such very heavy current loading, miniaturised monolithic Ti_4O_7 electrodes may not be truly corrosion resistant. There would be no advantage over titanium if the material required to be coated. The effectiveness of IR heating is dependent upon solution conductivity which, as far as natural waters are concerned, varies widely across the country, as does its hardness so that scale formation too, can be a problem with this technology. It is worth noting that a very substantial R&D effort into the design of such instant hot water boilers was made, with a succession of cushion-shaped electrodes. In such units, superheating occurs, and this can affect the chemistry and morphology of scale deposits which form.

4.2 ELECTRO OSMOTIC DAMP PROOFING

Few, if any, of the many techniques proposed to avoid rising damp are free from controversy, indeed the very phenomenon of rising damp itself remains controversial. This is as true of so-called "Electro-Osmotic Damp-Proofing" as it is of some other methods, such as Knapen Tubes. This or similar names, which imply the nature of the mechanism by which the method is claimed to work, were and perhaps still are, registered Trade Names. Electro-osmotic damp proofing is installed in buildings by drilling into the brickwork and inserting the individual anodes at regular intervals, above ground level. Each assembly embodies numerous anodes. The units are connected in series, and a current is passed between them and a cathode placed on the outside of the building. The theory is that, by electro-osmosis, water will migrate outwards from the (internal) anode to the (external) cathode, (Fig. 4-2.) thereby driving moisture from the walls to the ground outside.

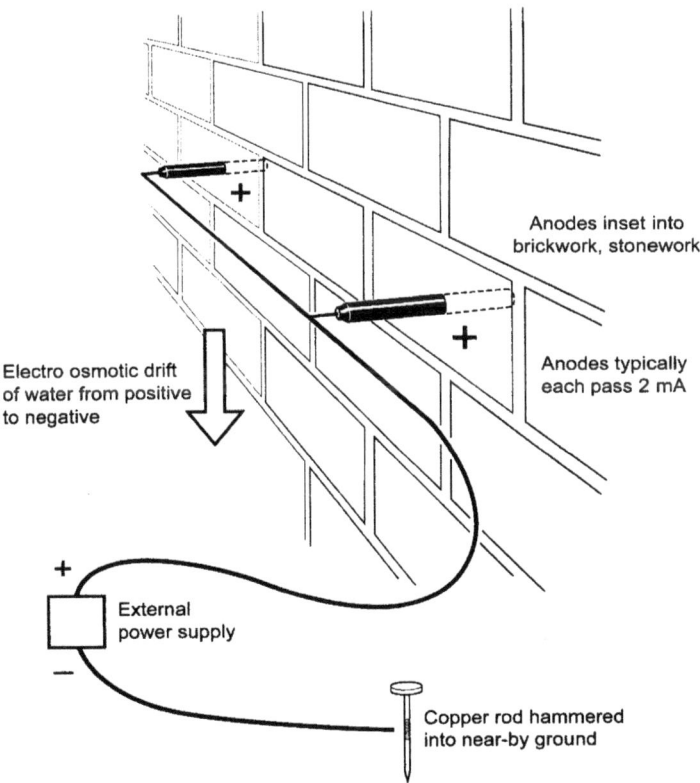

Anodes inset into
brickwork, stonework

Electro osmotic drift
of water from positive
to negative

Anodes typically
each pass 2 mA

External
power supply

Copper rod hammered
into near-by ground

Fig. 4-2. Electro Osmotic damp proofing of walls using uncoated, monolithic Ti_4O_7 ceramic electrodes grouted into brickwork.

As noted, most methods of damp-proofing are to some extent controversial, and electro-osmotic damp-proofing is no exception. The author is in no position either to express an opinion as to the comparative merits of the various methods, or indeed how efficacious any particular method might be. Iinjection of polymers and silicone treatments all have their protagonists. It has been argued that repairs to brickwork, plastering/rendering which usually accompany the installation of such devices, are probably more effective than any electro osmosis effect taking place under electrochemical polarisation. Be it as it may, in the UK at least, there are thousands of applications of the method to many types of building. Initially the anode used was a form of 'skip' plated platinum electroplated titanium wire i.e plated at intervals rather than uniformly. Where the wire was platinised it would be bent into a hair pin and then grouted into holes at appropriately spaced intervals, the titanium wire acting as the electrical supply to the platinised packs. In more recent years, platinised wire continues to be used, but is now cut from a reel and titanium circuitry completed by portable spot welding to a titanium wire ring main.

In theory lengths of monolithic Ti$_4$O$_7$ rod should be entirely satisfactory, electrical connection made by tightening titanium wire around prior to grouting into structure.

4.3 CHLORINE AND CHLORATE ELECTROLYSIS

Plant design associated with these important industrial electrochemical processes has continued to evolve, with one important change being the replacement of massive graphite anodes with noble/metal metal oxide-coated titanium anodes. Such electrodes are ideal in those technologies involving monopolar and multi finger monopolar designs. There has, however, always been a desire by some for more compact units based on filter press type bipolar construction. This raises a problem with titanium as cathodically generated hydrogen, especially at the elevated temperatures at which most chlorine and chlorate cells operate, will not only react with titanium to form a hydride at the surface, but also diffuse into the bulk titanium. A volume change results, eventually leading to shedding of the anode coating. For this reason, various types of hydrogen barrier have been conceived and tested, interposed between the coated titanium anode and the titanium cathode. (In most cell designs, such a barrier would also have to be electrically conductive). However the concept appears not to have been used so far on an industrial scale. Monolithic Ti$_4$O$_7$, size factor apart, provided it were non-porous, should allow the making of a stable bipolar electrode.

In the absence of commercially-available, dense, Ti$_4$O$_7$, porous material was variously coated with noble metal/anode electrocatalyst and tested in bipolar mode using equipment as sketched in Fig. 4-3. The electrolyte passing through the cell was saturated chlorinated brine at 95°C. Electrode pairs in the bipolar construction operated normally until a brine seepage occurred through the edges of the sample, indicating that even with coating it had not been possible to prevent some leakage paths occurring. The experiments nevertheless demonstrated that given availability of fully dense Ti_4O_7, then the diaphragm type chlorine cell construction is in theory a possibility.

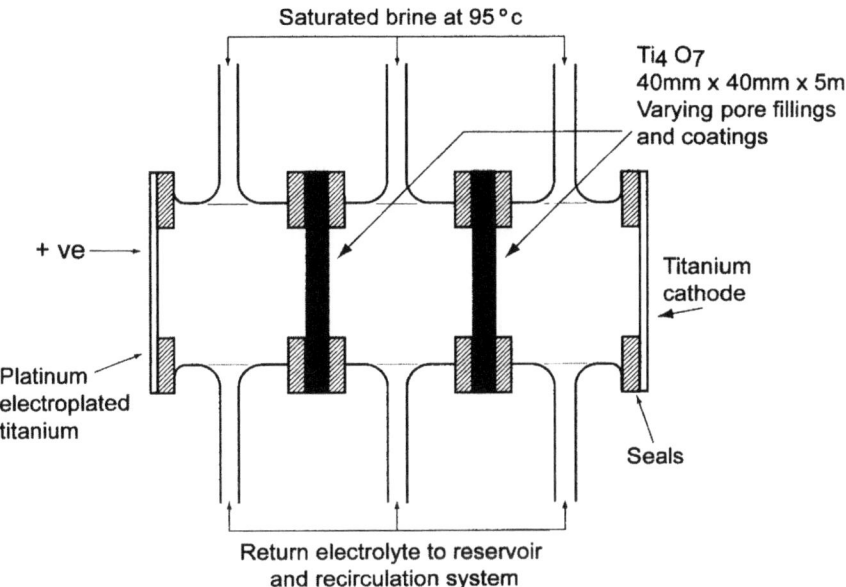

Fig. 4-3. Laboratory evaluation of variously coated monolithic Ti_4O_7 ceramic bipoles under chlor-alkali electrolysis conditions.

A golden opportunity for monolithic Ti_4O_7 in chlorate technology might have occurred had material been available a decade or more earlier. This is because plants existed with filter press type bipolar technology constructed from massive walls of graphite. With slight inefficiency in the chlorate process, and some oxygen evolved at the anode, the graphite suffered rapid wear. Plates of monolithic Ti_4O_7, noble metal oxide coated on the anode side, would have seemed an ideal replacement for the graphite. However, once the technology involving graphite had been replaced by the multi-finger monopolar technology using coated titanium, the high voltage rectifier systems needed for bipolar technology were discarded and hence the opportunity was lost.

4.4 HYDROCHLORIC ACID ELECTROLYSIS

With seemingly almost complete corrosion resistance towards hydrochloric acid, even when fully concentrated, Ti$_4$O$_7$ might seem the ideal material for anode substrates in this process. In practice, however, the process itself has lost importance in recent decades. The manufacture of chlorine by electrolysis of hydrochloric acid should be viewed as two quite separate markets. On the one hand, there is the large-plant market. Chemical manufacturers, usually with conventional chlor-alkali plants, use the chlorine on-site for chlorination of organics which can produce large quantities of HCl as an unwanted by-product. This would be fed to an electrolysis plant which produced chlorine and hydrogen, both were utilised on-site or pressurised, for external sale. In the UK, the process was not operated, or certainly not after the 1940's and ICI's Mond Division found alternative uses for the HCl. More recently, new chlorination reactions have been used, in which HCl is no longer formed as a by-product.

The second market is based on small-scale electrolysers for local generation of chlorine, to be used for swimming pool chlorination, or perhaps in power stations and other small-to-medium users of the gas. Electrolytic cells for HCl electrolysis are attractive in their simplicity. Unlike their brine electrolysis counterparts, they need no diaphragm or membrane. Significantly lower cell voltages mean that the energy costs per tonne of chlorine are two-thirds that required using brine-derived chlorine, or perhaps even less. The only serious problem with the classical electrolysers was the very high rate of graphite anode wear, which the new material overcomes. Against this, storage of aqueous HCl was, if not problematic, at least inconvenient.

It was calculated at one stage that a small electrolyser, capable of decomposing some 20 to 30 litres of concentrated hydrochloric acid would make all the chlorine necessary to treat a domestic swimming pool for a season. But there are risks in keeping so much acid, of the possibility of an accidental chlorine leak, and also in disposal of the hydrogen generated, and as far as is known, the concept has never been translated into hardware.

4.5 ELECTROLYTIC STERILISATION OF WATER

At one period, consideration was given to the use of monolithic Ti$_4$O$_7$ anodes in cells to sterilise water, and in particular, to destroy e-coli bacteria. It was held that destruction occurred by electrocution rather than electrochemical oxidation, but it was difficult to prove, since small concentrations of chloride are usually present in mains water, forming either hypochlorous acid or hypochlorite on electrolysis, and the chemical oxidation action of these species could not be discounted. In any case, the concept of "electrocution" when such a simple life-form as a bacterium is involved, is in any case dubious. In cells designed to carry out this process, electrolyte was passed over monolithic Ti_4O_7 having sharp corners, an optimum design being bars with a square cross-section. (see Fig. 4-4) This area of potential application could well merit further investigation.

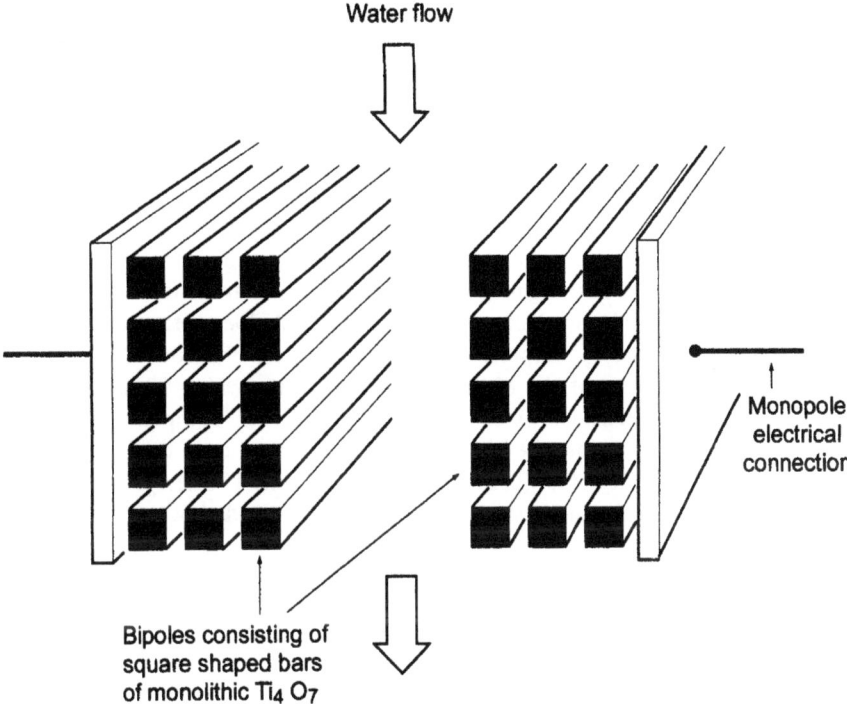

Fig. 4-4. Suggested design of cell to destroy e-coli bacteria in water (Courtesy of Dr Frank Goodridge, University of Newcastle-On-Tyne).

4.6   OTHER ELECTROCHEMICAL APPLICATIONS FOR MONOLITHIC Ti$_4$O$_7$

There are numerous possible electrochemical technologies where monolithic Ti$_4$O$_7$ might find application. There were initially great expectations for the material as an anode to promote redox reactions, the oxidation of anthraquinone, where trivalent chromium is electrochemically oxidised to form chromic acid. This reacts chemically with anthracene to form the quinone, and the resulting Cr(III) is recycled for re-oxidation. Known in the chemical industry as "indirect oxidation", the variable valency chromium compounds are known as the "mediator". The approach has been used for other industrial scale organic oxidations, and proposed for many more. Nor is chromium the only mediating species – cerium has found favour, and several other mediating species have been proposed. What is important to understand, is that the electrochemical reactions involved, are known as "redox" reactions and are in mechanistic terms, among the simplest of all electrochemical reactions, in some cases involving nothing more than the transfer of a single electron. The significance is that, at least in theory, nothing more is required of an electrode to carry out such a reaction, than that it be an electronic conductor which does not suffer chemical attack in the electrolyte. Expressed another way, an electrode for a redox reaction requires none of the electrocatalytic activity which is essential for chlorine evolving electrodes. Once again, Ti$_4$O$_7$ finds itself competing with lead (alloy) anodes and possibly the biggest obstacle to date lies in difficulty in the design and manufacture of large, flat sheet-form anodes. The design sketched in Fig. 2-29 might well be the basis for long-lasting electrode but has never been realised in practice.

4.7   MONOLITHIC Ti$_4$O$_7$ IN LIGHTWEIGHT BATTERIES

This application has the potential to constitute the largest single future use of monolithic Ti$_4$O$_7$. Battery technology, perhaps more than most other electrochemical technologies, is a complex interaction of chemical and electrochemical processes, manufacturing techniques and optimisation, where optimum cost and optimum performance may not coincide. The design and construction of any battery type is itself a complex matrix and the issue is compounded by recognising that numerous different battery types, i.e based on different electrochemical couples, compete with one another. Notwithstanding promising results from nickel-cadmium, sodium-sulphur and a number of other newer systems, the lead-acid battery system appears to have an assured future for many years to come. Which is not to suggest that there is no scope for further improvements in design and construction.

Those who are familiar with the lead-acid battery will be aware of its shortcomings. The active material ("paste") is retained in small windows formed in the cast lead "grid" which constitutes the actual battery plate (electrode). The grid serves two purposes, firstly to mechanically retain the active material, and secondly to act as electrical conductor, allowing charge to pass into or out of the plates. The basic design has remained virtually unchanged for over a century. Despite this, it remains true that lead is not among the best electrical conductors (among the metals); it is a soft material and mechanically weak, even when alloyed with antimony or other metals. And most crucially, it is a very heavy metal. Were an alternative material to be found, in which any one of the above weaknesses was mitigated, battery performance would immediately be transformed, in that the power-weight ratio or energy density (two rather different Figures of merit) would be increased. Such a substitution has one further huge attraction, in that it would allow existing battery manufacturing plant and machinery to continue to be used. In contrast to many examples cited above, this has the potential to be a "drop-in replacement" which could be introduced at minimal cost.

There is, however, an even more drastic option in re-engineering the classic lead-acid battery. As we know it today, a typical battery will be a case containing six completely separate compartments. Each compartment constitutes a 2V cell, and these are connected in series using lead interconnectors. There is no reason why this configuration could not be re-created using a series of bipolar electrodes (except for the first and last electrodes), and indeed such batteries have been constructed. There are further weight savings (since the current no longer has to be drawn from the entire length of a plate to the connector at the top, but can instead flow normally to the plane of the plate). There are savings in size too, in that the battery can effectively be "squashed up".

Lead-acid batteries serve a range of markets. Most competitive, without doubt, is the automotive market, where battery first-cost, as opposed to energy density etc, is of prime, perhaps sole importance. In other applications, cost is less the overriding issue, culminating with aerospace, where cost is almost irrelevant, performance and reliability being the dominant parameters. The 120 volt batteries used in aircraft are prime candidates for such development, using the ideas described above

How might the properties of Ti_4O_7 be harnessed in this technology ? One obvious way to reduce weight in a lead acid battery is to decrease the weight of individual plates. In theory a thin, impervious Ti_4O_7 grid would seem ideal, being capable of withstanding chemical attack from the concentrations of acid used. But as has been described in preceding sections, dense thin plates of monolithic Ti_4O_7 are not yet commercially available, and might possibly be too fragile. One route being pursued is to incorporate Ti_4O_7 particles in a hot setting resin. To maintain an adequate level of electrical conductivity, the particle loading must be high, certainly above 60%. Thus the practical aim is to make such plates non permeable to electrolyte seepage through their cross-section. Experimental batteries of this type are already being evaluated in portable electric tools.

Another possible use for Ti$_4$O$_7$ in lead-acid batteries would be to include particles of the material within the pastes used. In one sense, this would be replacing material capable of accepting or delivering electrical charge, with an inert substance. Against this, it would address an age-old problem which is that in its discharged state, the active lead/lead dioxide reverts to lead sulphate, which is an insulator. A heavily sulphated battery accepts charge very slowly, which in applications such as electric vehicles, is unacceptable. Incorporation of conductive particles of Ti$_4$O$_7$ could provide improved electrical conduction into the sulphated pastes, and so allow accelerated charging.[10]

5. Is Ti$_4$O$_7$ an Optimum Ceramic Electrode Choice?

5.1 SOME GENERAL REFLECTIONS

Technology has moved on apace since the relatively limited applications initially envisaged for non-porous Ti$_4$O$_7$ as a possible electrode material. Just as there are infinite variations in the metallurgical constitution of an alloy, so too does the science and technology of ceramics offer countless variants on a given theme. Innumerable solid state reactions are possible to allow formation of a host of complex materials, including those whose metastable state might offer novel properties and behaviour. Might some such materials be more electrically conducting than Ti$_4$O$_7$, or, still more exciting, exhibit superior electrocatalytic activity, or be cheaper? One can but speculate, but certainly, ceramics based on titanates and zirconates are well-known and indeed electrically conducting. A major attraction of Ti$_4$O$_7$ that it can be produced from TiO$_2$, a commodity chemical readily available across the world.

The technology of Ti$_4$O$_7$ manufacture will undoubtedly continue to evolve. New methods of forming may well be considered. Could material be made directly from the commonly occurring mineral ilmenite ? – probably not!. Might ammonia or carbon monoxide reductions be preferable to hydrogen ? Might solid-state reduction using carbon be a longer-term commercial option?

In an attempt to compare Ti$_4$O$_7$ with other similar solids, some preliminary experiments were carried out on titanium carbides and nitrides, both of which were found to have poorer corrosion resistance in sulphuric acid. Compounds exhibiting interesting electrochemical characteristics included TiSi$_2$ and ZrTiO$_4$.

(One might here mention that in the early days of fuel cell research, some of the above compounds, and many others, were screened for possible use as electrode materials. In addition, iron silicides have been used as anodes in CP applications). In some of the fuel cell studies, it was concluded that the surface of titanium carbides & nitrides degraded anodically to form TiO$_2$ layers on the surface.

It would be foolish in the extreme to suggest that no new electrically conducting materials with excellent corrosion resistance, will not be discovered. Interesting new materials will almost certainly be found. Such materials may find immediate application – or they might languish for many years, until an application is found. At the present time, however, it is fair to assert that in its combination of properties, monolithic Ti$_4$O$_7$ has no challengers. (see Tables 5-1 and 5-2).

TABLE 5-1

Durability of Titanium Carbide and Nitride in 1 M Sulphuric Acid

1) Titanium carbide was pressed from commercial powder and used as an anode in 1 M sulphuric acid at 100 A/m^2 current density. After 30 minutes the solution turned yellow, later deep yellow and later black. A peroxide test revealed high titanium in solution. Evidently TiC dissolves readily.

2) A titanium nitride tile was pressed from commercial purity powder and used as an anode in 1 M sulphuric acid at 100 A/m^2 current density. After 5 hours the solution was tested by peroxide and found to contain titanium ions. Evidently TiN dissolves readily in sulphuric acid.

3) A nitrided monolithic Ti$_4$O$_7$ tile was made anodic in 1 M sulphuric acid at 100 A/m^2 current density. After 3 days the bronze coloured TiN surface layer had disappeared, leaving the Ti$_4$O$_7$ passing current.

TABLE 5-2

**Reaction between titanium, titania and monolithic
Ti₄O₇ with nitrogen and ammonia**

Nitrogen	a) Commercial purity titanium sheet heated in nitrogen at 1200°C for 6 hours. Bronze coloured thin film of TiN formed; sheet embrittled.
	b) Nominally pure tile of TiO_2 heated for 5 hours at 1200°C. No TiN formed. By X-ray diffraction still TiO_2.
	c) Monolithic Ti_4O_7 heated for 6 hours at 1200°C. No TiN formed.
Ammonia	d) Titanium Dioxide powder heated for 3 hrs at 850°C. X-ray diffraction of darkened powder revealed TiO_2 + TiO/TiN. Chemical determination showed 44% TiN.
	e) Titanium dioxide powder heated 4 hrs at 1200°C. X-ray diffraction of darkened powder revealed TiN + trace TiO_2 + unidentified phase. Chemical determination showed 36% TiN.
	f) Nominally pure tile of TiO_2 heated for 3 hrs at 1000°C. A surface film of TiN formed. On grinding away, underlying material still TiO_2.
	g) Monolithic Ti_4O_7 heated for 6 hrs at 1000°C resulting in a surface film of bronze coloured TiN.
[A Japanese patent claimed heating TiO_2 powder in anatase form in ammonia resulted in a mixture of anatase (TiO_2), rutile (TiO_2) and TiO, with no mention of TiN. The above results suggest significant nitride is formed.]	

5.2 COSTS AND PRICES

All new materials must, in the same way as other manufactured products, seek to break out of the vicious circle where small production quantities imply high prices because the various economies of scale cannot be harnessed. Inherently, Ti$_4$O$_7$ should not be an expensive material. It is manufactured from titania, of which there is an abundance in the world's crust, and more importantly, which has for decades been a commodity product, used as a pigment in white paints, or to whiten paper. The conversion from titania is not particularly energy intensive. In the early days of the development in IMI, it was loosely predicted that once production reached reasonable levels, the cost would be around half that of titanium metal, a material with which it to some extent competes. At the time of writing, this is certainly not true, and "Ebonex" costs substantially more than titanium, whether the comparison is made on a weight basis, on a volume basis or using any other criterion. The reason lies mainly in the batch production methods at present used, and the costly equipment involved, e.g hot pressing. The development of alternative fabrication methods remains a tantalising possibility, perhaps using flame or plasma spraying, though such methods rarely produce materials at close to their theoretical densities. One recent patent (89) discloses such an approach.

6. Conclusions and Epilogue

6.1 SUCCESSES AND FAILURES

Monolithic Ti_4O_7 electrode material has certainly experienced a chequered development history. A commercial product is now available whose production is straightforward. Contrary to initial thinking, the most significant commercial application to date has not been associated with electrochlorination or electrowinning, but rather in different aspects of impressed current cathodic protection. Despite the corrosion resistance of bulk Ti_4O_7 in chloride solutions and hydrochloric acid, no large scale application in chlor-alkali technology has yet appeared. In spite of this, commercial applications for monolithic Ti_4O_7 which have been apparently dormant in the past 20 years do seem to be expanding. Each new commercial success offers an opportunity to lower the price, and thus, in a virtuous circle, attract even more potential uses.

6.2 EPILOGUE – A GATHERING OF MANY STRANDS

What has here been described, is an account of the development of a novel material, in the setting of an industrial laboratory. Industry tends to follow quite different paths from those likely to be followed in academic institutions. Industry goes to great lengths to ensure there is no prior art to cloud the issue, and will often expend much effort in patent searching and in obtaining patent protection. With the wisdom of hindsight, some of this effort might have been more wisely expended in learning more of the science and technology relating to, or peripheral to, the project.

In an era where computer databases make searching the literature a much simpler task than it was in the past, the probability of stumbling by chance on a new material might be expected to be small. In addition, and notwithstanding a comment above, solid-state science has progressed rapidly, and in considering perhaps 20 of the chemical elements as candidate building blocks, one is tempted to believe that the chances of identifying a new material would seem remote. The development of monolithic Ti$_4$O$_7$ ceramic, and its use to make electrodes was the result of ad-hoc experimentation, rather than design. A question one might put to a solid state scientist might be "Could one have predicted that Ti$_4$O$_7$ would possess the properties it has ?" With the passage of time, and the greater accessibility of scientific literature, the question has been asked "How novel is the concept of a monolithic Ti$_4$O$_7$ electrode?"

A publication from the Metal and Thermit Company in 1925[4] includes a description of the heating of rutile and carbon to 1500 to 2000°C to convert rutile into a dark blueish titanium suboxide. The development suggests the material will find ready application in fields that include electricity, electrochemistry and electrometallurgy, especially to form electrodes for use in situations where metallic conductors are found wanting, but it would seem such material never became commercial. A Foseco patent (86) concerned the use of such black oxide in the Electro slag or welding purposes. The General Electric Company in 1957 claimed conducting titanium oxide for cathode ray tubes as charge dissipating electrodes. A Mitsubishi patent of 1970 described various processes for deposition of titanium oxide films, for example by pyrolising in vacuo tetra-ethyl titanate, to form thin dielectric films. At the time of their filing, none of the inventors of the above-mentioned patents, and others of similar nature, had available to them the detailed description of the titanium-oxygen system shown in Fig. 2-1.

In around 1983, Mitsubishi Metal Corporation published information on a new electroconductive, non-toxic, titanium oxide pigment called titanium black, claiming tonnage quantities were being made each month. However the proposed use of this material was never made clear.

A Polish worker, Borowiec, published an article[11] in 1983 i.e. after the initial UK patent, on Ti$_4$O$_7$ as an electrode. Apparently development work had been proceeding in Poland in parallel with that in the UK.

Excimer laser printing of aircraft cables[8] might seem totally irrelevant to the present text, but it is clear that use of localised heating of TiO$_2$ pigment in cabling to form a lower black oxide to give permanent labelling, involves broadly similar chemistry.

In conclusion, it seems valid to suggest that the titanium-oxygen system has been a rich lode, both for theoreticians and for those in Applied Science, notably in the physics and chemistry of ceramics. This account describes how, perhaps against the odds, an electrically-conductive suboxide, with the potential to make a commercial electrode material was developed. As is so often the case, the author makes no attempt to conceal that an element of luck played a part in this. What else remains to be uncovered, not only in the titanium-oxygen system, but also in more complex systems such as the titanates, remains in the future for all that enormous research effort has already taken place in the field. In the process, much has been learnt about ceramic electrodes and how best to handle them in commercial application. The old ceramic adage, small is beautiful remains true. Optimum technology to produce fully dense monolithic Ti_4O_7 has yet to be introduced into commercial practice, and when it does the number of possible applications will greatly increase.

What might be termed the purely constitutional study of monolithic Ti_4O_7 has progressed little since the early days of the material. There are many parameters yet to be evaluated, especially those relating to its electrical properties. The effect of dopants, impurities and other minor additions to the Ti_4O_7 lattice, and thus on its properties have as yet received scant attention.

Beyond this, lie the related fields dealing with oxides of zirconium and perhaps silicon, and in respect of these, the author is happy to adopt the adage of the stockbroker – "leave something for the next man"

Acknowledgements

The author is conscious of being to some extent a narrator, recounting a saga in which many dedicated scientists and technologists were involved. It would be invidious to attempt to name them all, let alone to try to rank the extent of their contributions. Most of what has been described relates to work carried out in the Research and Development Department of IMI Ltd at Witton over the period 1975-1989. Much of the technology was then licensed to companies whose main aim was the exploitation of the material, including its manufacture and marketing. The author has tried to convey as accurate and balanced an account as possible, and hopes that as such, the exercise will be welcomed both by the directors of IMI plc and the newer companies to whom the baton has now been passed.

The author is also more than happy to acknowledge the Lambertville Ceramics Inc, of the USA, for their willingness to participate early on in the project, and likewise Messrs LCC, in Paris. He is conscious too, of the support provided by the community of ceramicists in the UK, from both academic and industrial organisations. He takes this opportunity of wishing Messrs Atraverda, who are now the main licensors of the technology, every success in the future. Their website (www.atraverda.com) is well worth visiting. Last but not least, he acknowledges the role of a long-standing friend, Anselm Kuhn, who after repeated exhortation, finally persuaded him that what is set down here was worth putting into the Public Domain.

REFERENCES AND FURTHER READING

Set out below, are references and further reading. All are numbered, and all include either title or a brief comment indicating the nature of the article. Numbering is partly for convenience, not all numbers are referred to in the text itself.

1. Nagasawa K., Kato Y., Bando Y. and Takada T., *J. Phys. Soc.* Japan (1970) **29**, 241 (Properties of Ti_4O_7 Single Crystals).
2. Smith J.R., Nahle A.H. and Walsh F.C., J. Applied Electrochemistry (1997) **27**, 813-820: (Scanning Probe Microscopy Studies of Ebonex Electrodes.)
3. Molchanova E.K; Glazunov S.G; "Phase diagrams of titanium alloys". Jerusalem IPST (1965). (transl. from Russian. No mention of Ti_4O_7 or Magneli series in Ti-O system) – *also:*
   Kubaschewski O; Kubaschewski-van Goldbeck, O; Rogl, P. & Franzen, H.F. Atomic Energy Review, Special Issue No. 9 publ, I.A.E.C. Vienna (1983) "Titanium. Physico-Chemical Properties of its Compounds & Alloys" (pp. 22-30 "Thermochemical properties of Ti-O system". pp. 129-134 "Ti-O Equilibrium Diagrams").
4. UK Patent 232,680 (1925). (Metal & Thermit Corp.) Improvements in the Production of a Form of Titanium Oxide (Electrically Conducting).
5. Graves J.E., PhD thesis University of Southampton (1991) (Electrochemistry of Titanium Oxide Ceramic Electrodes.)
6. Miller R.R., MSc Thesis; Wake Forest University North Carolina (1987) (Electrochemical Properties of a Non-Stoichiometric Titanium Oxide Material, Ebonex.)
7. Mitsubishi American Metal Market, 26th Aug. (1983) (New Titanium Oxide Pigment Developed. "An electroconductive and non-toxic titanium pigment, called Titanium Black, has been developed and marketed by Mitsubishi Metal Corporation. Derived from the reduction of titanium dioxide".)
8. Williams S.W., (BAe Sowerby Research Centre) private communication. (Excimer Laser Printing of Aircraft Cables.)
9. Graves J.E., Pletcher D., Clarke R.L. and Walsh F.C., *J. Applied Electrochemistry* (1991) **21** 848-857 (Electrochemistry of Magneli phase titanium oxide ceramic electrodes. Part 1. The deposition and properties of metal coatings.)
10. Kao W., Patel P. and Haberichter, *J. Electrochem. Soc.,* (1997) **144** (No. 6) (1907-1910) (Formation Enhancement of a Lead/Acid Battery Positive Plate by Barium Metaplumbate and Ebonex®).
11. Borowiec K; "Nonstoicheometric titanium oxides as an electrode material" p. 31 in: "Terkel Rosenqvist Symp. In honour of Terkel Rosenqvist". Norwegian Inst. Technol. Trondheim, May 8-11, 1988. Ed. S E Olsen & J Kr. Tuset. Publ. Norges tekniske Hoegskole, Div'n of Metallurgy, (1988).
12. Geraghty K.G. and Donaghey L.F., Thin Solid Films (1977) **40** 375-383 (Preparation of Suboxides in the Ti-O system by Reactive Sputtering.)

13. Baumard J.F., Panis D. and Anthony A.M. *J. Solid State Chem.* (1977) **20** 43-51 (Study of Ti-O System between Ti$_3$O$_5$ and TiO$_2$ at High Temperature by means of Electrical Resistivity.)

14. Clark C.H.W., Turner R.B. and Powell D.G., Trans. British Ceramic Society (1961) **60** (No. 5) 330-342 (Properties of Conducting Glaze Based on TiO$_2$)

15. Wood G.J., Bursill L.A., Yoshida K. and Yamada Y., *Phil. Mag. A.* (1982) **46** (No. 1) 75-86 (Oxidation Mechanism of the Crystallographic Shear phase Ti$_4$O$_7$)

16. Houlihan J.F. and Mulay L.N., *Phy. Stat. Sol. B* (1974) **61** 647-657 (Electronic Properties and Defect Structure of Ti$_4$O$_7$: Correlation of Magnetic Susceptibility, Electrical Conductivity, and Structural Parameters using EPR Spectroscopy.)

17. Schlenker C. and Buder R., Proceedings of the International Conf. On Ferrites held in Japan (1980) Sept/Oct. p. 123-130 (Metal-Insulator Transitions in Ti$_4$O$_7$ and (Ti$_{(1-x)}$-V$_x$)$_4$O$_7$ Experiments and theories.)

18. Marezio M., McWhan D.B., Dernier P.D. and Remeika J.P., Physical Review Letters (1973) **28** (No. 21) 1390-1393 (Charge Localization at Metal-Insulator Transitions in Ti$_4$O$_7$ and V$_4$O$_7$.)

19. Murray J.L; (Editor) ASM Monograph Series (1987), ISBN 0871702487 (Phase diagrams of Binary Titanium Alloys.)

20. Farndon E.E., Pletcher D. and Saraby-Reintjes A., *Electrochemica Acta* **42** 1269-1279 (Electrodeposition of platinum onto a conducting ceramic Ebonex®)

21. Garcia E.A., Metaux Corrosion-Industrie (1978) 639-640 357-382 (Relation between oxide film formation and oxygen absorption between 650-950°C for titanium.)

22. Garcia E.A., Metaux Corrosion Industrie (1978) **53** No. 638 329-331) (as above, but different technique – also includes diagram of the titanium-oxygen equilibrium system.)

23. Kendall K., Birchall J.D., and Alford N. McN, paper 7 (in) "Adv. ceramics in chemical process engineering" Meeting Papers, Ed. B C H Steele & D B Thompson. Inst. of Ceramics. British Ceramic Proceedings No. 43. (1988) ISSN 02684373 (Properties & Applications of Electrode Materials based on a new process for the Manufacture of Ti$_4$O$_7$ – Ebonex®)

24. Hayfield P.C.S. and Clarke R.L. Electrochemical Society Spring Meeting, Los Angeles CA. May 7-12 (1989) (Electrochemical Characteristics and uses of Magneli Phase Titanium Oxide Ceramic Electrodes.)

25. Böder H. Herbst H. and Rubisch O., Chem. Ing. Techn. Vol. 49 (4) 331-333 (sintered anodes for Chloralkali-Elektrolysis based on Titanium Suboxide)

26. Hauffe K., "Oxydation von Metallen und Metallegierungen" (1965) Plenum Press.

27. Kubachewchewski O., and Hopkins B.E.,"Oxidation of Metals & Alloys" Butterworths, London 1962.

28. European Patent 0047 595: (IMI Marston Exelsior Ltd) (1981) (Electrode material, electrode and electrochemical Cell. This was so-called "IMI Master-Patent on Ti$_4$O$_7$ with US 4,422,917 (1983) as equivalent).

29. US 4,912,286 (Ebonex Technologies Inc). (1990) (Use of a titanium suboxide as a corrosion resisting heating element.)

30. European Patent 081155 (Mitsubishi Kinzoku Kaboshiki Kaisha) (1982) (Process for producing an electrically conducting blue to black powder by heating titanium dioxide powder in an atmosphere of ammonia gas at temperatures from 500 to 950°C.)

31. US 2,369,266 (American Lava Corporation) (1945) (Electrically conductive Ceramic Threadguide.)

32. UK Pat 817,233 (General Electric Company) (1959) (Method of Rendering Titanium Dioxide Films Electrically Conductive.)

33. US 4,931,213 (Cass R.B.) (1990) (Electrically-Conductive Titanium Suboxides. Reduction using intercalated graphite.)

34. UK Pat 1,373,214 (Lord F.W.,) (1971) (Composite Materials based on polymer + 4 different sizes of powder to improve packing density).

35. European Pat 360,942 (Ebonex Technologies Inc.) (Conductive Ceramic Support for active material in a lead-acid storage battery)

36. Hayfield, P C S; Platinum Metals Rev. (1998) (Pts 1-3).Vol. **42**, 27-33, 46-55, 116-122 ("Development of the Noble Metal/Oxide Coated Titanium Electrode")

37. Anderson, S., and Jahnberg, L., Arkiv f. Kemi 1963, **23**, (No. 39) 413-425 (Crystal Structure Studies on the Homologous Series $Ti_nO_{(2n-1)}$, $V_nO_{(2n-1)}$ and $Ti_{(x-2)}Cr_2O_{(2n-1)}$).

38. Beckenbridge, R.C., and Hasler W.R., Physical Review 1953, **91**, 793-801 (Electrical Properties of Titanium Dioxide Semiconductors.)

39. Pearson, A.D., *J. Phys Chem Solids* 1958, **8**, 316-327 (Studies of the Lower Oxides of Titanium)

40. Anderson, S., Sundholm, A., and Magneli, A., *Acta Chemica Scandinavica* 1959, **13**, 987-997 (A Homologous Series of Mixed Titanium Chromium Oxides $Ti_{(n-2)}Cr_2O_{(2n-1)}$ Isomorphous with the series $Ti_nO_{(2n-1)}$ and $V_nO_{(2n-1)}$)

41. Honig, J.M., and Read, T.E., Physical Review 1968, **114**, 1030-1027 (Electrical Properties of Ti_2O_3 Single Crystals)

42. Houlihan J.F., and Mulay L.N., Mat. Research Bull. 1976, **6**, 737-742 (Characterisation of and Electronic Structural Studies on the Oxides of Titanium EPR Linewidths)

43. Roy, R., and White, W.B., J. of Crystal Growth 1972, **13/14**, 78-83 (Growth of Titanium Oxide Crystals of Controlled Stoichiometry)

44. Marezio, M., McWhan, D.B., Dermier, P.D., and Remeika J.D., Physical Review Letters 1972, **28**, (21), 1390-1393 (Charge Localization at Metal/Insulator Transitions in Ti_4O_7 and V_2O_7)

45. Schlenker, C., Buder, R., Schlenka M., Houlihan, J.F., and McLay, L.N., *Phys. State Sol.* 1972, **54**, 247-252 (Electron Paramagnetic Resonance Spectra of Ti_3O_5 Crystals.)

46. Houlihan J.F., Dissertation Abstracts, 1974, **35**, (6) (EPR Spectroscopy of Mixed Valence Oxides of Titanium and their Electronic Transitions)

47. Houlihan, J.F., and Mulay, L.N., Inorganic Chemistry, 1974, **13**, (3) 745-747 (Electron Paramagnetic Resonance Spectra of Oxides of Titanium. Effective Magnetic Moments of Ti^{3+} Ions.)

48. Houlihan, J.F., and Roy, R., J. Amer. Ceramic Society – Discussions and Notes; May, 1974, 234-235 (Lattice Energies of Ti$_n$O$_{2n-1}$ Phases and the Related Homologs V$_n$O$_{2n-1}$ and Ti$_{2n}$ Cr$_2$O$_{2n-1}$ Calculated from Crystal-Chemical Parameters.)

49. Houlihan, J.F., Dowley, W.J., and Mulay, L.N., J. of Solid State Chemistry, 1975, **12**, 265-269 (Magnetic Susceptibility and EPR Spectra of Titanium Oxides: Correlation of Magnetic Parameters with Transport Properties and Composition)

50. Grossklaus, W; Bunshah R.F; J. Vac. Sci. Technol. 1975, **12**, (1) 593-7 (Synthesis of Various Oxides in the Ti-O system by Reactive Evaporation and Activated Reactive Evaporation Techniques.)

51. Houlihan, J.P., and Mulay, L.N., *Phys. Stat. Sol.* 1974, **61B**, 647-657 (Electronic Properties and Defect Structure of Ti$_4$O$_7$: Correlation of Magnetic Susceptibility, Electrical Conductivity and Structural Parameters via EPR Spectroscopy)

52. Schlenker, C., Lakkis, S., and Coey, J.M.D., Physical Review Letters 1974, **32**, (23), 1318-1321 (Heat Capacity and Metal-Insulator Transitions in Ti$_4$O$_7$ single Crystals.)

53. Houlihan, F.F., and Mulay, L.N., Phys. Stat. Sol. 1974, **65B**, 513-519 (Correlation of Magnetic Susceptibility, Electrical Conductivity, and Structural Parameters of Ti$_3$O$_5$ via EPR Spectroscopy)

54. Lorenz, B., and Thiele, D., Phys. Stat. Sol. 1976, 77B, K 177-K 181 (Model of the Semiconductor – Semiconductor Transitions in Ti$_3$O$_7$)

55. Lakkis, S., Schlenker, C., Chakraverty, B.K., and Buder, R., Physical Review, B, 1976, **14**, (4), 1429-1440 (Metal Insulator Transitions in Ti$_4$O$_7$ Single Crystals: Crystal Characterisation, Specific Heat, and Electron Paramagnetic Resonance.)

56. Houlihan, J.F., and Madacsi, D.P., Mat. Res. Bull, 1976, II, 307-314 (Crystal Field Splitting in Ti$_3$O$_5$ Via EPT Spectroscopy)

57. Kaplan, D., Schlenker, C., and Since, J.J., Philosophical Magazine, 1977, **36**, (5), 1275-1279 (Optical Absorption and Metal-Insulator Transitions in Ti$_4$O$_7$)

58. Bartholemew, R.F., and Frankl, D.R., Physical Review, 1969, **157**, (3), 828-833 (Electrical Properties of Some Titanium Oxides)

59. Garcia, E.A., Metaux Corros. Ind. (1978) Vol. 53, 329-331 (Relation Entre Les Mechanismes de Croissance du Film d'Oxyde et de la Dissolution de l'Oxygene dans le Metal Sous-jacent au Cours de l'Oxydation du Titane entre 650 et 950°C) (Relationship between oxide film growth mechanism & oxygen dissolution in underlying metal, on oxidation of Ti)

60. Fazekas, D., Friend, R.H., and Marseglia, E.A., Phil. Mag. B. 1980, **42**, (3) 479-484 (Model for the Impurity Induced Stabilization of the Intermediate Phase in Ti$_4$O$_7$)

61. Strobel, P., and Le Page, Y., J. of Crystal Growth 1982, 56, 723-726 (Growth of Ti$_9$O$_{17}$ Crystals by Chemical Vapor Transport.)

62. Bando, Y., Muraraka, S., Shimada, Y., Kyoto, M., and Takada, T.J. of Crystal Growth, 1981, **53**, 443-450 (Crystal Growth of Ti$_n$O$_{2n-1}$ with Large n by Chemical Transport Reaction.)

63. Sam-hyo-hong Acta. Chem. Scand. – 1982, **A36**, (3), 201-217 (Crystal Growth of Some Intermediate Titanium Oxide Phases & Ti$_4$O$_7$ and Ti$_2$O$_3$ by Chemical Transport Reactions.)

64. Fairhurst, S.A., Inglis, A.D., Le Page, Y., Morton, J.R., and Preston, K.F., Chemical Phys. Letters 1983, **95**, (45) 444-448 (EPR Spectrum of Ti^{7+}_2 in Single Crystals of Ti_6O_{11})

65. LePage, Y., and Strobel P., J. of Solid State Chemistry, 1983, **47**, 6-15 (Structural Chemistry of Magneli Phases Ti_nO_{2n-1} (4<n>9) III. Valence Ordering of Titanium in Ti_6O_{11} at 130°K.)

66. Strobel, P., and LePage, Y., J. of Materials Science 1982, **17**, 2424-2430 (Crystal Growth of Ti_nO_{2n-1} Oxides (n = 2 to 9))

67. Inglis, A.D., LePage, Y., Strobel, P., and Hurd, C.M., J. Phys. G. Solid State Phys., 1983, **16**, 317-333 (Electrical Conductance of Crystalline Ti_nO_{2n-1} for n = 4-9)

68. Kaito, C., Iwanishi, M., Harada, T., Miyano, T., and Shiojiri, M., Transactions of the Japanese Institute of Metals, 1983, **24**, (6), 450-460 (High Resolution Electron Microscopic Studies of Crystallographic Shear Structures in Reduced Rutile Crystals.)

69. Fairhurst, S.A., Inglis, A.D., LePage, Y., Morton, J.R., and Preston, K.F., J. of Magnetic Resonance 1983, **54**, 300-304 (Intrinsic Paramagnetic Dimers in the Magnetic Titanium Oxides.)

70. Miyano, T., Iwanishi, M., Harada, T., Kaito, C., and Shiojiri, M., Oguk. Mag. A 1983, **48**, (1) 163-167 (A (110) CS Structure in Reduced Rutile Crystals.)

71. Pollock, R.J., Houlihan, J.F., Bain, A.N., and Coryea, B.S., Met. Res. Bull. 1984, **19**, 17-24 (Electrochemical Properties of a New Electrode Material, Ti_4O_7)

72. Frearujssibm E., and Carlsson, Jan-Otto., Thin Solid Films 1985, **124**, 109-16 (Chemical Vapour Deposition of Titanium Oxides in the Composition Range $TiO_{1.60}$-$TiO_{1.75}$)

73. Goldschmidt, D., and Watanabe, M., Mat. Res. Bull. 1985, **20**, 65-70 (X-Ray Diffraction of Polycrystalline Ti_4O_7)

74. MacChesney J B & Muan A, American Mineralogist (1959) Vol. 44, 925-945. and (1961) Vol. 46, 572-82 (Studies of the system iron-oxide & titanium oxide & phase equilbria at liquidus temperatures & low oxygen pressures).

75. Goodenough, J.B. "Metallic oxides" pp. 344-51 (in) "Solid state chemistry of energy conversion & storage" Symp. Sponsored by Div'n of Inorganic Chemistry, 171st ACS Mtg New York, April 5-8, 1976.

76. Improvements in sintered hard compositions GB 768532; 768329 (General Electric) Contains 15-20% Cr, bal. Ti suboxides, used for cutting tools, drawing dies etc.

77. Rogl, P., Franzen, H.F., Editor (Komarek K.L., Ed.) Atomic Energy Review Special Issue No. 9 (International Atomic Agency Vienna 1983) (in): (Titanium Oxygen System)

78. Marczio, M., McWhan, D.B., Dernice, P.D., and Remeika, J.P., Phys. Review Letters 1972, **28**, (21), 1390-1391 McN.Charge Localisation at Metal-Insulator Transitions in Ti_4O_7)

79. Breckenbridge, R.G., and Hosler, W.R., Phys. Review 1953, **91**, (4), 793-802 (Electric Properties of Titanium Dioxide Semiconductors.)

80. Kendall, K., Birchall, J.D., Alford, N., McN., and Clarke, R.L., Inst. Ceram Proc., 1989, **43**, 131-138 (Properties and Applications of Electrode Materials based on a new Process for Manufacture of Ti_4O_7 (Ebonex®))

81. European Patent Publications. No. 183453, (1986) Kendall, K., Alford, N. McN. Kendall, K., Clegg, W.J., and Birchall, J.D., Howard, A.J., Kendall, K., and Raistrick, J.H., (ICI Publications and Patents Relating to use of Polymer addition and shear in fabrication.)

82. Alford, N., McN. Birchall, J D., and Kendall, K., Nature 1987, 330, **51**

83. Dutoit, E.C., Cardon, F., Vanden Kerchove, F., and Gomes, W.P., J. Appl. Electrochem. 1978, **8**, 247-252 (Comparative study of electrochemical and photoelectrochemical reactions at Semiconducting Oxidized Ti and TiO_2 single crystal electrodes in view of energy conversion applications.)

84. Hayfield, P.C.S., and Hill, A., Int. J. of Restoration of Buildings and Monuments, 2000, **6**, (6), 647-654 (Ebonex® Discrete Anodes in the Cathodic Protection of Rebars in concrete: Performance of Ebofix Grouts;)

85. Andreeva, V.V., Corrosion 1964, **20**, (2), 35t-46t. (Behaviour and Nature of Thin Oxide Films on some Metals in Gaseous Media and in Electrolyte Solutions (includes titanium).

86. BP 1,219,580 (Foseco International Ltd) (1971) (Titanium oxides in electroslag process)

87. BP 1,021,583 A1 (Thermal Sprayed Electrodes)

88. Electrochemical Method & Electrode US 6120675 (Atraverda) Hollow porous tube electrode made of Ti suboxide with electrical connection.

89. PS 100 00 979 (DaimlerChrysler) (2000) (flame & plasma spraying of titanium Suboxide coatings.) See also:
    Storz, O & Gasthuber H; (DaimlerChrysler) "Tribological properties of thermal sprayed Magneli type coatings with different titanium:oxygen stoichiometries" Surf. & Coating Technol. (2001), Vol. **140** (2), 76-81

90. Nafion-bonded porous titanium oxide electrodes for oxygen evolution: towards a regenerative fuel cell. Hamnett A.V., Stevens, P.S. and Wingate, R.D., J. Appl. Electrochem., 1991, 21 (11), 982-985

91. Electrolytic Oxidation of trichloroethylene using a ceramic anode. Chen, G., Betterton, E.A. and Arnold, R.G., J. Appl. Electrochemistry, 1999, 29, 961-970.

92. Sol-gel film-preparation of Novel Electrodes for the Electrocatalytic Oxidation of Organic Pollutants in water. Grimm, J., Bessarabor, D., Maier, W., Storek, S. and Sanderson, R.D., Desalination, 1998, 115 (3), 295-302.

93. Studies of Platinised Ebonex Electrodes Farndon, E.E. and Pletcher, D., Electrochim Acta, 1997, 42 , 1281

94. Electrodes based on Magneli Phase Titanium Oxides: The Properties and Applications of Ebonex® Materials. Smith, J., Walsh, F.C. and Clarke, R.L., J. Applied Electrochem., 1998, 28, 1021-1033

95. Preparation of Various titanium suboxide powders by reduction of TiO_2 with silicon Hauf, C., Kniep, R. and Pfaff, G., J. Materials Science, 1999, 34, 1287-1292

96. Characteristics of the conducting ceramics TiO and Ebonex Ti$_4$O$_7$ as electrode materials, Park, S-Y, Mho, S, *et al*; Thin Solid Films, 1995, Vol. 258, 5-9

97. Electrochemical treatment of human wastes in a packed bed reactor. Tennakoon, C.L.K., Bhardwaj, R.C., Bockris, J. O'M., J. Appl. Electrochem., 1996, 26 (1), 18-29 (includes use of SnO$_2$/Sb$_2$O$_3$ coated Ebonex electrodes.)

98. Electrochemical incineration of garden wastes. Lin, G.H., Chen. S., Tennakoon, C.L.K., Bhardwaj, R.C., Leong, S.M., Mervwa, R. and Bockris, J.O'M., Int. Forum Electrolysis Chem. Ind., 9th 1995, 431-438

99. The Reduction of Oxygen on titanium oxide electrodes. Baez, V.B., Graves, J.E., and Pletcher, D., J. Electroanal Chem., 1992, 340 (1-2), 273-86

100. A new fluoride Resistant ceramic electrode for electrochemical effluent treatment processes Harnsberger, S.K. and Romoda, I., Plat. Surf. Finish, 1990 77 (7), 40-42

101. Electron transfer reactions at Ebonex ceramic electrodes Miller-Folk, R.R., Noftle, R.E., Pletcher, D., J. Electrochem. Interfacial Electrochem., 1989, 274 (1-2), 257-261

102. The electrochemistry of Magneli phase titanium oxide ceramic electrodes Part II Ozone generation at Ebonex and Ebonex/lead dioxide anodes. Graves, J.E., Pletcher, D., Clarke, R.L. and Walsh, F.C., J. Appl. Electrochem. 1992, 22 (3), 200-203.

103. Production of ethene oxides in a sieve-plate electrochemical reactor. Pt 1. Influence of sieve plate design, electrode material & pH. Scott K, Hui W, J. Appl., Electrochem, 1996, 26 (1) 10-17 (includes use of iridium oxide coated Ebonex electrodes)

104. Cathodic corrosion protection of steel reinforced concrete structures with a new composite conductive paint. Schwartz W; Int'l J. for Restoration of Buildings & Monuments (abstract at www.ijr.ethz.ch/abstracts_6_00.html)

105. Electrochemical properties of a new electrode material, Ti$_4$O$_7$. Pollock, R J; Houlihan, J F; *et al*; Mater. Res. Bull. (1984), Vol. 19, 17-24. Cyclic voltammograms of catalysed & uncatalysed material (Ru or Pt), incl. oxygen & hydrogen evolution. ESR results show presence of small amounts of Magneli phases.

106. Electrical conductivity of Ebonex. Coux, M. *et al*. Proc. 6th European Lead Battery Conf. Publ. in. J. Power Sources (1999), Vol. 78 (1/2) 115-122

107. Zador S; (in) Inst. Min, Met Symp. "Electromotive Force Measurements in High Temperature Systems" (1968) Ed. Alcock, G.B; 145-150 "Non-stoicheometric measurements in dioxides of the rutile structure" (paper was cited as prior art in opposing patent applications).

OTHER PATENTS

108. Electrokinetic Treatment of Contaminated Soil Sunderland, J.G. and Roberts, E., (EA Technology Ltd) 1998, DE 694 05 403 T2 (uses Ebonex anodes)

109. Electrochemical Cell for Recovery of Metal Ions from Dilute Solutions Sunderland, J.G., and Dalrymple, I., (EA Technology Ltd) 1999, DE 694 15 027 T2 (uses Ebonex anodes)

110. Electrode, Means for its Manufacture and Applications for it DE 3432652 C2 (Tillmetz W; Gnann M, Wabner D;)

111. Manufacture of titanium suboxide Horikawa, Matsuhide, Kagohashi & Wataru JP 91124693 (includes Ti$_3$O$_5$ as a coloured fibre pigment).

112. Production of conductive titanium suboxide particulates for electrochemical cells Clarke, R.L., PCT Int. Appl. No. WO 9214683, 1992

113. Manufacture of titanium suboxide articles. Hill, A., UK Pat. Appl. GB 228174 Al, 1995 (Atraverda Ltd) (non wicking article produced by combined high temperature and pressure processing. See also DE 694 19 178 T2., publ. 1999).

114. Layered composites consisting of an inorganic substrate (including Ti$_4$O$_7$), an organo-silane bonding interlayer and polymeric top coating, and their manufacture Mai, S. and Henns, P., EP 95107201

115. Electrically conductive composition and use thereof Moreland, P.J., EP 44329 Al; 1991 (Shaped conductive article comprises organic polymeric material and particulate electrically conducting titanium suboxide).

116. Manufacture of electrically conductive Zirconia ceramics Horinochi, K., Muri, M., Kameda, I. JP 8944096 (1989) (Ceramic comprises 60-90 vol% ZrO$_2$ + 10 -40 vol % TiN made by mixing ZrO$_2$ powder with titanium suboxide (eg TiO) moulding and sintering in nitrogen.)

117. Electrical conductors formed of sub-oxides of titanium. Clarke, R.L., US 88232954, 1988 (includes a corrosion resistant heater)

118. Electrochemical Cell Brooks, N., Eur. Pat. Appl. EP 443230 1991.

119. Ceramic electrode material based on titanium oxides. Polish Pat PL 153305 B1 Przyluski, J; et al; (1991) (Chem Abs. 116;243821)

120. Electrochemical cell containing titanium suboxide electrode. US 5281496 (Atraverda)

121. Conductive titanium suboxide particulates. US 5173215 (Atraverda); also EP 0572559 & AU 1555692

122. Bronze-grey glazing film & window with same US 4861680 (Southwall Technologies Inc). Reactive sputtered film.

123. Transparent, colorless, electrically conductive coating US 4194022 (PPG Industries) Vacuum deposited coating for windows etc.

124. Method of preparing electron tube incl. sputtering titianium suboxide on dielectric components thereof US 3309302 (Varian Associates)

125. Alkali metal compounds of suboxide of titanium & derivatives therefrom. US 1731364

126. Method of producing titanium suboxide articles. US 5733489 (Hill, A.). Binding Magneli phase powder & heat-treating to form corrosion-resistant non-wicking article.

127. Electrically conductive coatings GB 2028376 (PPG). Vacuum dep'n to form optically clear, colorless multilayer coating incl. Ti suboxides.

128. Light-absorbing & anti-reflective coating for sunglasses US 5729323 (Bausch & Lomb) Multilayer coating incl. Ti suboxide

129. Pane of transparent material having low emissivity US 5962115 (Leybold) Multilayer coating incl. Ti suboxide

130. Electrochemical cell & process EP 0443230 (Ebonex Techn.) Porous gas cathode made of Ti suboxide.

131. Titanium Suboxide Powder JP 8059240 (Mitsubishi Materials). Method of manufacture.

132. Production of titanium suboxide JP 5009028; 5004818; 4349121; 4342420 ; 4325416; 1290529; 4317413 (Toho Titanium). Vacuum process for production of Ti_3O_8.

133. Conductive coating of titanium suboxide GB 2309230 (Atraverda)

134. Current Collecting Elements US 5521029 (AT & T) Substrate with corrosion protection using Ti suboxide coating.

135. Electrode WO 0102626 (Atraverda) Elongate hollow tube made of porous Ti suboxide.

Websites and Miscellaneous Information from the Web

The internet offers a useful source of information which partly complements conventional scientific and patent literature, and partly includes abstracts of the former. Using search terms such as "Ebonex" or "Titanium Suboxides" with one of the main search engines such as www.google.com yields brief mentions of published papers, the websites of commercial organisations using the material or selling it and other references to it. Some of the commercial websites include technical information. In some instances particular individuals or University departments list their publications, and sometimes on-going research. Information on patents is now readily available on the internet.

Atraverda Ltd – Main licensors of Ebonex®. An overview of the material and its applications, still under construction at the time of writing.

    www.atraverda.com

Fosroc Ltd – Ebonex anodes for discrete cathodic protection of reinforced concrete structure and steel framed buildings.

    www.fosroc.com

C-Probe – Corrosion protection of structures and buildings using Ebonex anodes. Website provides illustrated case histories.

    www.c-probe.com

Geokinetics Australia Ltd. (a wholly owned subsidiary of the Moltoni Group) Electro remediation of Impurities in Ground Waters.

    www.geokinetics.com

APEX Technical Report RC41 Improved anodes for use in electrowinning.
www.apexis.co.uk/fitdocs/summaries/RC41S.htm.

American University of Beirut

www.aub.edu.lb/aub-online/research/22report/as/chemistry.htm

cites projects such as:

Nahle, A., The study of Ebonex electrodes for the simultaneous determination of Cd., Pb and Cu ions by anodic stripping voltammetry.

Smith J R, Nahle A; Walsh F C; Scanning probe microscopy studies of Ebonex electrodes (AFM & STM used to study topography of porous & fully-hardened Ebonex, with AFM used to study early stages of copper electrodeposition onto the material.

Ebonex used in a lithium-manganese battery mentioned at:

www.oorg.kemi.uu.se/kurser/oomn2/Main/Projekt.htm

Electrolytic cell for the inactivation of viruses in water, publ. Foundation for Water Research. DW10743. £25.00 from FWR, Allen House, The Listons, Liston Road, Marlow, Bucks SLY 1FD Uses bipolar rod cell with Ebonex anodes. Summary at:

www.fwr.org/waterq.dwi0743.htm

Oxidation of iodine anions in aqueous solutions at catalysed & non-catalysed Ebonex electrodes. Pjescici, M; Draskovic-Boskovici, I; *et al*; (University of Montenegro).

http://208.5.159.10/meetings/future/200/abstracts/symposia/e1//0718.pdf (from google search for "ebonex")

Pjescici, M., Draskovic – Boskovici, I.; Menpusz, S., Biagojevici, n. and Komnenici, V. (Universities of Montenegro & Belgrade)

http://208.5.159.10/meetings/future/200/abstracts/symposia/e1/0718.pdf)

Index

Aluminium, in titania 4
Ammonia, reaction of Ti oxides with 79
Analyses, spectrographic, of titania 6
Anode design, large 50
Anode materials, options for 1
Anodes, for industrial electrochemical processes 2,12
Anodes, point, for cathodic protection, 62
Applications of Ebonex ™, miscellaneous 49, 74, 82

Bactericidal action 73
Band-gap, of Ti oxides 32
Batteries, electrodes for 25, 74
Binders 5, 24
Bipolar technology 55, 71,
Birchall, J.D, ceramics studies 23
Black surface film 10,12
Brittleness, of ceramics 24

Cables, excimer laser printing of 82
Cathodic protection 58 *et seq.* 65, 66
Cathodic protection, of concrete rebars 59
Cathodic protection, anodes for 53
Cathodic protection, Ebonex™ tiles and 63
Chlor-alkali electrolysis 71
Chlorate electrolysis, anodes 70
Chlorine electrolysis, anodes 70
Choice, optimum ceramic 77
Coating, of Ti suboxides 31
Colour, of Ti oxides 32
Corrosion resistance, of Ebonex™ 36 *et seq.* 39

Costs, of Ebonex™ 80
Current Reversal, see Periodic Current Reversal

Damp-proofing, electro-osmotic 69
Density, of titania 6
Density, of titania, after H red'n 17
Dichtol, sealant material 29

Ebonex™, commercial successes & failures 82
Ebonex™, precursor developments 82
Ebonex ™, Trademark 2
Electrical conductivity, oxide composition and 38
Electrical connections, to Ebonex™ 44 *et seq,* 56
Electrical resistivity, meas't of 20
Electrical resistivity, of Ti suboxide 32, 33, 34
Electrical resistivity, of Ti suboxide-PVDF mixts. 26
Electrical resistivity, purity effect of 36
Electrochemical cell design, for Ebonex™ 46
Electrochemical properties, of Ebonex™ 36
Electrochlorinators 51
Electrode boiling 67
Electrode potentials, in brine 44
Electrowinning 1, 2, 3, 4
Electrowinning, large anode design for 50
Ellipsometry, for surface studies 9, 12
Equilibrium diagram, Ti-O 10 *et seq.*
Etching, of Ti, on oxalic acid 22

Flame spraying, of Ti oxides 80
Furnace, for reduction of titania 14

History, of Ebonex™ development 1, 82
Hydride, Ti 52
Hydrochloric acid, electrolysis of 72
Hydrofluoric acid 2
Hydrofluoric acid, corrosion of Ti and
    Ebonex™ 38, 40
Hydrogen uptake, by various Ti oxides
    58
Hydrogen reduction, of titania 8, 15, 16,
    18
Hypochorite, electrolytic 52

IMI 1
Imperial Metal Industries Ltd, *see* IMI
Intermetallics, Ti-based 2

Lead-acid batteries, electrodes for 74
    et seq.
Lead-silver alloy anodes 2

Machining, of Ebonex™ 46 *et seq.*
Magneli series, oxides 4, 13, 14
Manganese, electrowinning 4
Manufacture of Ti suboxides, options
    for 13
Marston Palmer Ltd 2
Melting point, of Ti oxides 33
Monopolar technology 71

Nitrogen, reaction of Ti oxides with 79

Oxidation, alloying effects on thermal 43
Oxidation, of Ti suboxides, in air 30, 32

Paint, Ebonex™ powder loaded 60, 61
Periodic current reversal, effect on
    Ebonex™ 42, 52, 53, 54
Phase diagram, see equilibrium diagram
Plasma spraying, of Ti oxides 80
Plating, of Ti suboxide 31 *et seq.*

Platinum, as catalyst for titania red'n 21
Polarisation plots, effect of alloying 43
Polarisation plots, for Ti suboxide &
    metals etc 41
Polarity reversal, effect of 42
Pore filling, of Ebonex™ 28
Pore structure, of monolithic oxide 27
Porosity, of Ti suboxides 25, 26, 64
Pressing, uniaxial hot 23, 24
Prices, of Ebonex™ 80
Properties, of misc. oxides, nitrides,
    carbides & metals 35

Resistive heating, of liquids, 68

Scale formation, in brine 51, 54, 56
Scale formation, and resistive heating 68
Seawater, behaviour in 29
Sintering aids 5
Solid state reaction, of Ti oxides etc 22
Specific heat, of Ti oxides 32
Sterilisation, of water 72
Stoicheometry, from weight loss of
    titania 17
Structure, of Ti oxides 11, 13
Swimming pool, electrochlorinators
    for 51, 57

Textile machinery 3
Thermal conductivity, of Ti oxides 32
Thermal expansion, coefficient of 32
Thread guides, textile 3
Tiles, weight variation in Ebonex™
    63, 64
Titania, blued 3, 4, 7
Titanium carbide, behaviour of 78
Titanium-copper intermetallics 2
Titanium dioxide, capacitors 3
Titanium, intermetallic cmpds 2
Titanium nitride, behaviour of 78
Titanium oxalate, auto-reduction of 21
Titanium oxalate, formation in Ti
    etching 22
Titanium oxide film, properties 3

Vitrification, of ceramics 5, 24   X-ray diffraction diagram 11

Wadsley defects 5, 13   Zinc, electrowinning 4, 49
Warping, of titania tiles on red'n 18
Water, sterilisation of 72
Weight loss, of titania as guide to
    composition 17, 19

Books from the
Royal Society of Chemistry

The RSC publishes an extensive range of high quality
books dealing with all aspects of chemical science, including:

- **Analytical chemistry**
- **Biological sciences**
- **Food chemistry**
- **Environmental science and technology**
- **Medicinal and pharmaceutical chemistry**
- **Polymer and materials science**

From specialist titles to books with wide appeal, for audiences
ranging from academics and industrialists to students and tutors,
the RSC offers the complete range at very competitive prices.

For further details, including information
for prospective authors visit:

www.rsc.org/books

RSC ADS-21010222-MONO

www.ingramcontent.com/pod-product-compliance
Ingram Content Group UK Ltd.
Pitfield, Milton Keynes, MK11 3LW, UK
UKHW021817190226
468218UK00011B/27